How to grow
your own
NUTS

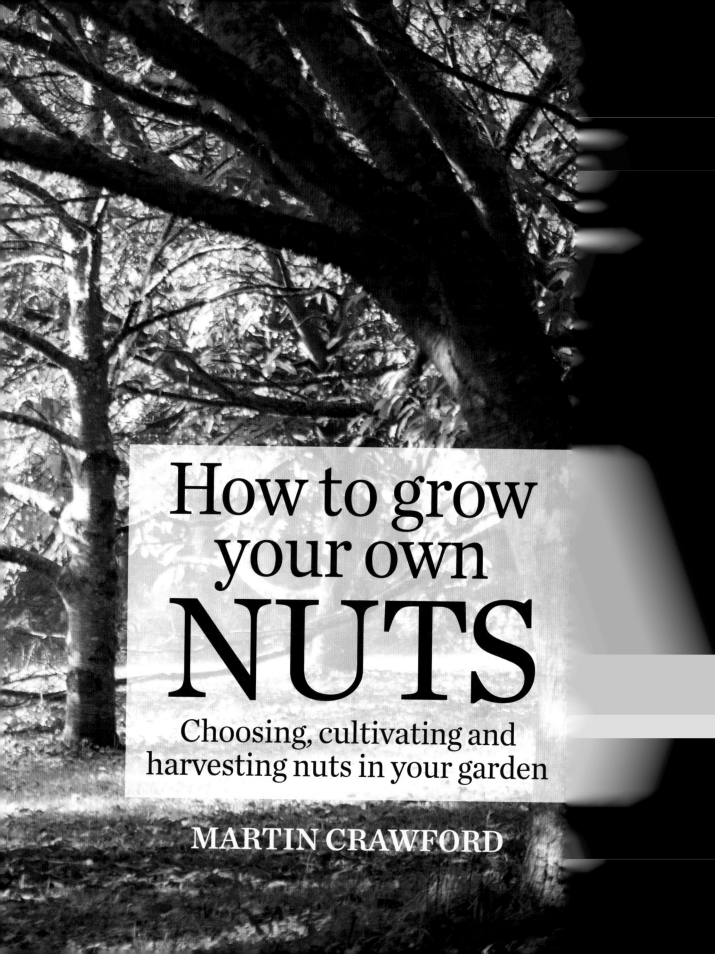

How to grow your own
NUTS

Choosing, cultivating and harvesting nuts in your garden

MARTIN CRAWFORD

Published by
Green Books
An imprint of UIT Cambridge Ltd
www.greenbooks.co.uk

PO Box 145, Cambridge CB4 1GQ, England
+44 (0)1223 302 041

First published in 2016, in England
Reprinted 2021

Interior illustrations © 2016, 2021 Marion Smylie-Wild
Front cover photography © 2016, 2021 Joanna Brown
Credit for interior photographs are on page 312.
Interior photographs not listed on page 312 are by the author.
Design by Jayne Jones

ISBN: 978 0 85784 393 7 (hardback)
ISBN: 978 0 85784 552 8 (paperback)
ISBN: 978 0 85784 394 4 (ebook)

ep-1-2

Contents

Introduction

You've probably eaten nuts for most of your life, as most people have, but perhaps assumed that they are difficult to grow or that they require a special climate to thrive. Maybe you've been growing your own fruit and vegetables for a while and are wondering what it takes to grow nuts? Or you've increasingly heard how healthy it is to eat nuts? Well, this book is designed to help you discover what nuts are available, and how to grow, harvest and process them so that you can cultivate your own delicious and bountiful crops.

What is a nut?

A nut is a hard-shelled fruit containing an edible kernel that is a powerhouse of nutrients. The shell can be thick and tough, as with walnut or hickory nuts, or leathery like sweet chestnuts. The kernel is surrounded by a thin skin called the pellicle, which is sometimes bitter when fresh and becomes papery when dry. The whole nut itself is borne within a burr, husk or other structure – prickly sweet chestnut burrs are a well-known example. The burr or husk may split to release the nut at ripening (as with walnuts), or the whole structure may fall to the ground intact (as with heartnuts).

Why grow your own nuts?

There are many good reasons for cultivating nut trees.

Nutritious and flavourful food

Since prehistory, nut trees have been used as a source of food as well as other materials. Indeed your ancient ancestors are likely to have depended on nuts as one of their staple foods for thousands of years. Centuries of deliberate cultivation and selection are documented for plants such as walnut, and these have led to the thin-shelled nuts we know today. However, there has been little selection for species such as hickories, and the thick-shelled nuts are nearer to their wild state.

Nuts are one of the most nutritious foods available, so it is little wonder they remain a valued food today. High in calories, nuts complement other foods including fruit and most vegetables. Some nuts are high in carbohydrates, like sweet chestnuts, and are in effect a 'tree grain'. Others, like walnuts, contain abundant healthy oils, which can be extracted for use in cooking.

Nuts are also full of minerals, vitamins and antioxidants, and in recent years there's been a significant amount of medical research showing the health benefits of eating oily nuts like walnuts. In fact we may all be healthier if we ate less carbohydrates and more healthy oils such as those in nuts. The Australian 'Nuts for Life' programme has recently prompted health officials to include oily nuts on the recommended list of daily foods, and there is no doubt that the health benefits of nuts will soon become better known everywhere. In 2015 new research[1] has shown that

* eating 56g (2oz) of walnuts per day improves blood vessel wall function and cholesterol levels, reducing the risk of heart disease;
* regular nut consumption is associated with reduced mortality, reduces the incidence of diabetes and helps glycaemic control in those with type 2 diabetes.

As well as nutritional benefits, there's also flavour to consider. Anybody who grows some of their own food knows how much better it tastes when picked fresh from the garden than bought from a supermarket, and the same goes for nuts. Local, home-grown nuts are generally much better quality than shop-bought imported varieties, and their superior flavour may surprise you! Just like fruit varieties, many nut varieties have been selected and flavour is one attribute they have been selected for, so cultivated nuts often taste (and crop) better than wild collected ones.

Low maintenance

Growing tree crops, whether fruits or nuts, always involves less work than growing low-growing vegetables or arable crops. You only have to plant trees once. After they are established, they don't suffer terrible competition from weeds, they can tolerate droughts and short floods, and usually only need minor maintenance in addition to harvesting the crop.

Sustainability

Like most shrub- or tree-based systems, growing nuts is inherently more sustainable than cultivating carbohydrate crops such as cereals or oil crops such as oilseed rape, because nut crops require low energy inputs. Most importantly, a tree crop system does not require soil cultivation (or at most only between the tree rows), which is one of the major factors making annual-based agriculture less sustainable in energy terms.

1 See Njike, V. Y. et al. (2015). 'Walnut ingestion in adults at risk for diabetes: Effects on body composition, diet quality, and cardiac risk measures'. *BMJ Open Diabetes Research & Care*, 3(1): e000115. DOI: 10.1136/bmjdrc-2015-000115; Ros, E. (2015). 'Nuts and CVD'. *British Journal of Nutrition*, 113(supplement S2): 111-20. DOI: http://dx.doi.org/10.1017/S0007114514003924; Del Gobbo, L. C. et al. (2015). 'Effects of tree nuts on blood lipids, apolipoproteins, and blood pressure: Systematic review, meta-analysis, and dose-response of 61 controlled intervention trials'. *American Journal of Clinical Nutrition*, 102(6): 1347-1356. DOI: 10.3945/ajcn.115.110965; 'PREDIMED: A five year Mediterranean and mixed nuts diet study from Spain'. www.nutsforlife.com.au/wp-content/uploads/2015/09/Nuts-Predimed-brochure_2015_single-pages.pdf

Nuts as multipurpose trees

Some nut trees are grown for other uses apart from their nut harvest.

Timber. Hickory, sweet chestnut and walnut are especially valued for their timber. If timber production is a long-term aim, then trees should be trained with a minimum of 3m (10') of clear trunk. Using nurse trees can help achieve good straight form. There can be a trade-off here, though, as timber trees grow straightest when packed closely together, whereas widely spaced trees are best for nut production.

Sap. Tree sap can be tapped from the hickory and walnut families, just like sap is tapped from birches and maples, and be used in the same ways to make syrup and wine, etc. Tapping will slow down tree growth and can only start when trees are over 20cm (8") in diameter.

Edible mushrooms. If any trees are coppiced or pollarded, the branches can be used to grow edible mushrooms such as shiitake.

Honey. Almond and sweet chestnut orchards are highly valued by beekeepers, who usually move hives into them prior to flowering.

Dyes. The husks of some trees (e.g. walnut family) and bark of others (e.g. oak) have long been used as dye and are still used commercially.

Medicinals. Ginkgo leaves are well known for increasing blood flow in the brain. Walnut leaves also have a long history of being used for their fungicidal properties.

Examples of many other uses are given in the species listing in Part II.

The more food we can get from trees or shrubs, the more resilient our food growing will be, not just with respect to pests and diseases but also to climate change. With climate, what damages plants isn't so much gradually rising temperatures but extreme climatic events – floods, droughts, hurricanes, and so on. These extreme events are becoming more frequent, and the plants most susceptible to damage are annuals with their weak root systems. Trees and shrubs are much more tolerant.

Suitable for gardens and commercial orchards

Nut plants vary from medium-size shrubs to large trees, and there is something to fit any size of site, climate, location and growing method.

In a small garden, bladdernut and yellowhorn shrubs can be grown as well as dwarf walnuts, while in a larger space, sweet chestnuts and walnuts become possible, among others. On a smaller scale, problems of pests and diseases

tend to be much less, as smaller planting systems tend to have more diversity and more resilience.

If you're looking for a commercially viable crop, nuts lend themselves well to commercial growing systems. The trees need little work, and harvesting is nearly always from the ground after the nuts fall and so is fairly easily mechanized. Nut growing can be a profitable business. However, nuts, just like any other crop, can be prone to all sorts of pests and diseases when grown in monocultures, so many commercial systems often use herbicides, insecticides, etc., which seriously restricts their sustainability. More sustainable commercial systems use mixed species in plantings and organic cultivation methods.

There is great interest in agroforestry plantings using nut trees too, with EU funding available in some countries. There's potential for growing nut trees both on a larger scale with arable crops and grazing pastures and in smaller-scale, more complex systems like forest gardens. Some of these interplanting systems are described in Chapter 1.

Aesthetic appeal

You'll find that nut trees can be very beautiful and ornamental as well as productive. Species such as almonds are attractive in flower; others, like ginkgos and walnuts, in time make stately trees that will last for many generations. Growing trees can be highly satisfying and give you other, less tangible, benefits too.

The scope of this book

This book is designed to give you all the information you need to cultivate your own nuts, whether your aim is to grow one or two nut trees in your garden or you're interested in cultivating nut crops commercially. Part I will introduce you to the types of nut you can grow and explain how to plan for them, grow and process them. Part II is an A to Z of nuts that gives more detailed information about cultivation, choosing suitable cultivars and the diverse uses of nuts. Appendix 1 contains nutritional tables for all the nuts featured in the book.

The book covers nuts that can be grown in the temperate and continental climates of Europe, North America, Australia and New Zealand, though much of it will also be applicable to the rest of the world. Tropical nuts such as macadamia are not covered here, but they are included in the Resources section.

How to use this book

Wherever possible I have used both Latin and common names for nut and other plant species, and these names are listed in Appendix 2.

The A to Z species listing gives varied amounts of information, according to what is available at the time of writing. While there is good information about flowering times, etc. for some nut species – for example walnuts – there is often less information for other, less common nuts. I have included cultivar tables for more common species, but there may not be any cultivars for uncommon nuts like yellowhorn. Also, where there is no relevant information about flowering, for example, that category is not included in the cultivar table.

Each nut described in the A to Z listing is accompanied by a hardiness rating. This indicates the minimum average winter temperatures that a plant can tolerate.

Hardiness systems

Two different systems are used to give a rating of plant hardiness, and I have employed both. The first is the USDA hardiness zone system that is widely used in Continental Europe and North America. The second is the more recent hardiness rating system created by the Royal Horticultural Society (RHS). This is better suited to temperate maritime conditions (Britain, Ireland, etc.), where the great variability of winter temperatures and the cooler summers mean the USDA system does not work so well.

USDA hardiness zones

In the USDA system, zones are defined as areas with a particular average minimum temperature within a 10°F range. Initially defined as 10 zones (1 to 10), this has recently been expanded to 14 zones and also includes subdivisions of each numbered zone (zone 5a, etc.), each of which covers a 5°F range. The USDA has published a zone map for the USA using this new scale at www.planthardiness.ars.usda.gov However, for most other parts of the world the original 10-zone system is still used. The zones are listed with equivalent RHS ratings in the table below.

Maps showing USDA hardiness zones in the USA, Canada, Europe, Australia and Japan are also available online, for example at http://jelitto.com/Plant-Information/Plant-Hardiess-Zones

These maps can be used as a general guide to which plants will survive where you are. How-

Zone number	Average min temp (°C)	Average min temp (°F)	Equivalent RHS rating
1	-51 to -46	-60 to -50	H7
2	-46 to -40	-50 to -40	H7
3	-40 to -34	-40 to -30	H7
4	-34 to -29	-30 to -20	H7
5	-29 to -23	-20 to -10	H7
6	-23 to -18	-10 to 0	H6 to H7
7	-18 to -12	0 to 10	H5 to H6
8	-12 to -7	10 to 20	H4 to H5
9	-7 to -1	20 to 30	H3 to H4
10	-1 to 4	30 to 40	H2 to H3

ever, in urban areas, near buildings and on a sheltered southern hillside you may 'gain' a whole hardiness zone; whereas in hollows, valleys and northern hillsides you may 'lose' a zone of hardiness. Plants near rivers or lakes / large ponds that don't freeze benefit from the extra warmth in winter, worth up to half a zone of extra hardiness.

Plant hardiness is also affected by seasonal conditions in a particular year – a dry autumn leads to better hardening-off of growth and greater winter hardiness than a mild, wet autumn.

RHS hardiness ratings

The RHS system (see table below) also uses minimum winter temperature scales, but takes into account temperature swings, especially in spring and autumn, which are common in temperate maritime regions. (These cause the common damage to new growth associated with late spring frosts, and in autumn the problem of new growth being cut back by frosts before it ripens properly.)

The ratings are H1 to H7 (H1 and its sub-ratings are not included here, as all are very tender).

Metric and imperial values

Values throughout this book are given in metric, with imperial conversions for lengths, weights, temperature and area.

Rating	Average min temp (°C)	Average min temp (°F)	Category	Definition
H2	1 to 5	34 to 41	Tender	Tolerant of low temperatures, but will not survive being frozen. Except in frost-free inner-city areas or coastal extremities, requires glasshouse conditions in winter, but can be grown outdoors once risk of frost is over.
H3	-5 to 1	23 to 34	Half hardy – mild winter	Hardy in coastal/mild areas, except in hard winters and at risk from sudden (early) frosts. May be hardy elsewhere with wall shelter or good microclimate. Likely to be damaged or killed in cold winters, particularly with no snow cover or if potted. Can survive with artificial winter protection.
H4	-10 to -5	14 to 23	Hardy – average winter	Hardy through most of the UK apart from inland valleys, at altitude and central/northerly locations. May suffer foliage damage and stem dieback in harsh winters in cold gardens. Some normally hardy plants may die in long, wet winters in heavy or poorly drained soil. Plants in pots are more vulnerable.
H5	-15 to -10	5 to 14	Hardy – cold winter	Hardy in most places throughout the UK even in severe winters. May not withstand open/exposed sites or central/northern locations. Many evergreens suffer foliage damage, and plants in pots will be at increased risk.
H6	-20 to -15	-4 to 5	Hardy – very cold winter	Hardy across the UK and northern Europe. Many plants grown in containers will be damaged unless given protection.
H7	< -20	< -4	Very hardy	Hardy in the severest European continental climates.

Part ONE

Cultivating & processing nuts

Chapter ONE

Growing nut trees

There are many different types of nut tree you can grow – they range from shrubs to large trees, and some like particular soils, while others need cross-pollinating with another tree, and so on. Characteristics of your site – especially the space you have available and the soil type – are likely to determine what is possible. But you might also have particular preferences or be a potential commercial grower looking for a profitable crop to grow.

Once you have narrowed down the list of nuts you want to grow, you then need to decide on a planting arrangement before sourcing your trees and planting them out. This chapter explains the ideal conditions for growing nut trees. It also describes pollination requirements and planting scheme options as well as factors to consider when selecting and planting nut trees.

Walnut affected by walnut blight disease.

Choosing nut trees

Which nut tree species you can cultivate is determined by several factors:

❋ Your soil type and soil acidity. Some trees will grow on all soils, while others have more specific requirements (see 'Soils', page 26).

❋ How sheltered your site is. If you don't have shelter, you should grow some (see 'Shelter', page 20). On an exposed site smaller trees are more likely to succeed.

❋ How long you want to wait for your first crops. Some trees, like almonds, crop in 3 years; others, like monkey puzzle, take 25-30 years! Are you planting for yourself or also for your children?

❋ The space you have available, as some nut trees become large with age. The trees in this book can be divided into three categories:

– **Small** (under 5m/16') – bladdernut, chinkapin, dwarf walnut, yellowhorn
– **Medium** (5-10m/16-33') – almond, hazelnut, pecan (in cool climates), trazel
– **Large** (10-25m/33-80') – black walnut, buartnut, butternut, ginkgo, heartnut, hickory, monkey puzzle, oak, pecan (Southern Europe/USA), pine nut, sweet chestnut, walnut

As well as choosing which species to grow, you will also need to select cultivars (if there are any), and there's detailed information in Part II to help you do this.

Site characteristics

Various characteristics of your site, whether a garden or field, will favour some nut species and be unfavourable to others, and this should be a major factor in choosing what to grow. Trying to grow something very unsuitable for the site will be a waste of your time and effort, although there are some factors you can alter (like shelter) and some that are difficult or impossible to alter (like aspect).

Size of site

Not all nut trees grow as large as sweet chestnut and walnut trees. Certainly, these large trees need a garden or site at least 200m² (2,150 ft²) and for more than one tree double that or more.

Smaller nut trees include almond, hazel, dwarf walnut and yellowhorn. Almonds grow the same size as fruit trees like plums so you don't need that much space – you can even fan train them flat against a wall. Hazels need about 25m² (27 ft²) each, whereas yellowhorn, a much smaller shrub, will fit in most gardens.

Aspect

Most nut trees described in this book require full sun for a complete crop. Many of the trees themselves can tolerate light shade and a few quite deep shade (see table right). However, cropping is reduced in shade and will be virtually zero in full deciduous shade. Bladdernuts are the most shade-tolerant, and these will crop well in deciduous shade. In this section and throughout the book I talk about south aspects, but of course northern aspects are desirable in the southern hemisphere.

Nut trees tolerating shade	
Crop well in shade	Partial crop in light shade *
Bladdernut	Chinkapin
	Hazelnut

'Light shade' here means still several hours of sun per day.

Since most species demand full sun, nut trees should not be sited too close to other large trees (whether in a hedge or some other tree on your land) to the south or west particularly, and to an extent also to the east. Sun from the south and west (midday and afternoon sun) is more valuable to plants than easterly sun (morning sun), because air temperatures are usually higher after midday and photosynthesis works more efficiently.

If you have an existing hedge or windbreak to the south or west, or are planting one (see 'Shelter' below), then how close to the hedge you can plant your nut trees depends on the eventual size of the hedge and the nut trees. In general, the closest it is safe to plant near a hedge is about two-thirds (66 per cent) of the maximum height of the hedge. So, for example, if you have a hedge that is 6m (20') high, plant no closer than 4m (13') from the hedge. If you plant too close, then your nut trees are likely to lean away from the hedge and become unbalanced as they grow. In time they are more likely to fall over, as well as possibly being excessively shaded.

Frost pockets are also best avoided. A frost pocket is an area where cold air gets trapped on cold frosty nights and cannot escape by flowing downhill – typically found just above a hedge placed along the contour on a slope. The air in frost pockets can be significantly colder on frosty nights, so young plants there can be dam-

I can't emphasize enough how important shelter is for nut trees.

aged, particularly in spring. Older plants tend to grow out of the frost pocket so the problem reduces over time. If you need to plant in a frost pocket, make sure the trees are hardy and if possible use later-leafing cultivars. Walnuts, for example, can suffer from late frosts but there are plenty of late-leafing selections to choose.

Shelter

I can't emphasize enough how important shelter is for nut trees. Most nut trees are wind-pollinated (see 'Pollination', page 29), and strong winds will blow pollen out of the area before it can go from tree to tree. A few nut trees are bee-pollinated and these require shelter for the bees to fly at maximum potential. Some flower very early in spring, so evergreen or dense deciduous shelter may be appropriate for these.

If you don't have good shelter, then you should plant windbreak hedges to provide it. Ideally use tree or shrub species that grow only to the height you need and so won't need work to restrict their height.

A single windbreak hedge slows the wind very significantly for a distance of about 7 or 8 times the height of the hedge (see Figure 1). However, usually you will have a second hedge on the other side of your growing area (Figure 2). Initially, when you have just started planting small trees, this extends the area of good shelter to up to 20 times the height of the hedges. But as your nut trees grow, the wind is gradually deflected higher and higher over the whole area (Figure 3).

So try to choose species for your windbreak hedges that are as high or slightly higher than your nut trees will grow. Also have a second hedge no further than about 20 times the maximum hedge height. If you have a large nut orchard, this may mean a series of parallel hedges.

There are many different trees and shrubs that can be used in windbreak hedges – you might know some that do very well in your locality. A few of my favourites that are quite adaptable to many regions are listed in the table on page 22 – also see Figure 4. Many of these are nitrogen-fixing plants that will help supply nitrogen (see Chapter 2, page 46) to your nut trees. Some are good fruiting plants, although fruiting on the outer exposed side of the hedge is likely to be limited.

Windbreak hedges can have secondary functions as well, including helping accumulate other minerals, bee plants, shelter for wildlife, and a source of logs or poles from coppicing, etc.

A single line of trees/shrubs is enough in most windbreaks, unless you have an exceptionally

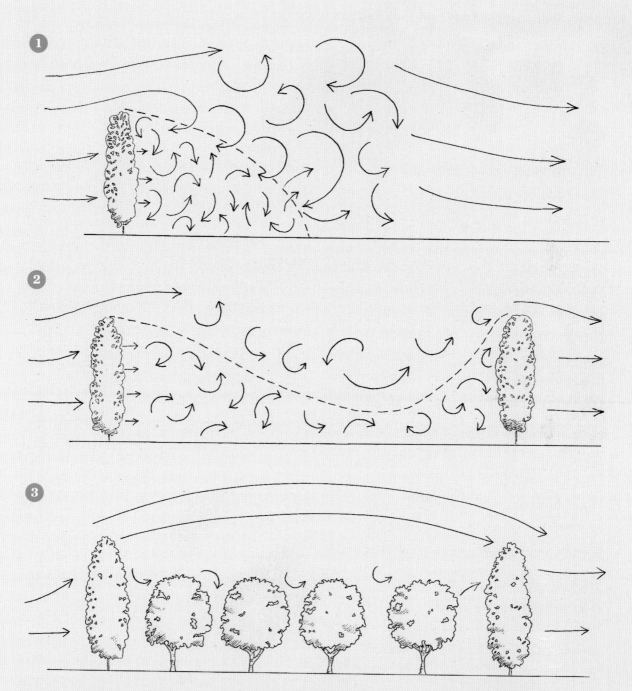

Figure 1 Wind flow through a windbreak. Figure 2 The protected area lengthens with a second hedge.
Figure 3 When trees are grown the wind flows over the top.

Windbreak trees and shrubs

Small (under 4m/13')		
Alnus sinuata	Sitka alder	Nitrogen-fixing
Alnus viridis	Green alder	Nitrogen-fixing
Berberis koreana	Korean barberry	Good edible fruit
Caragana arborescens	Siberian pea tree	Nitrogen-fixing, best in continental climate
Cytisus scoparius	Broom	Nitrogen-fixing
Elaeagnus multiflora	Goumi	Nitrogen-fixing, good edible fruit
Elaeagnus umbellata	Autumn olive	Nitrogen-fixing, good edible fruit
Elaeagnus x ebbingei		Nitrogen-fixing, edible fruit, evergreen
Hippophae rhamnoides	Sea buckthorn	Nitrogen-fixing, good edible fruit
Medium (4-8m/13-26')		
Crataegus spp.	Hawthorns	Species with good edible fruit
Elaeagnus umbellata	Autumn olive	Nitrogen-fixing, good edible fruit
Prunus cerasifera	Cherry plum	Good edible fruit
Prunus insititia	Damson	Edible fruit
Salix spp.	Willows	
Large (8m/26' plus)		
Acer pseudoplatanus	Sycamore	
Alnus cordata	Italian alder	Nitrogen-fixing
Alnus glutinosa	Alder	Nitrogen-fixing
Alnus rubra	Red alder	Nitrogen-fixing
Chamaecyparis lawsoniana	Lawson cypress	Evergreen
Cupressus macrocarpa	Monterey cypress	Evergreen
Pinus spp.	Pines	Evergreen
Prunus cerasifera	Cherry plum	Good edible fruit

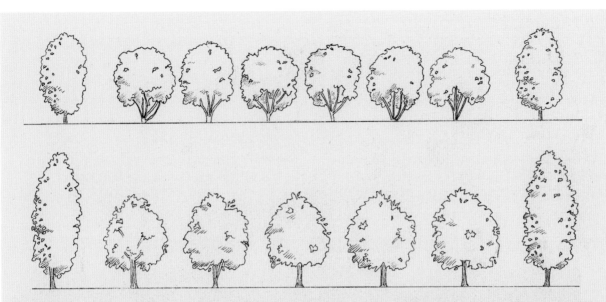

Figure 4 Examples of windbreaks. Above – cherry plum windbreak with hazel trees; below – Italian alder windbreak with walnut trees.

Figure 5 A double-row windbreak can be useful in extremely exposed locations, where the extra width needed gives much greater protection.

exposed site. Plant quite close in the line – typically 50-150cm (1'8"-5') apart, depending on the size of the trees. And remember to mulch well to suppress weeds, as young trees hate competition, especially from grasses.

If you are planting a line of trees that might get thin down below in future years (some trees have a habit of dropping their lower branches over time), then consider planting a second line of something shrubby to fill the gap that might evolve (see Figure 5). Planting the shrubbier plant on the *inside* of the tree line results in slightly better shelter than planting it on the outside (with the latter there is more of an aerofoil shape to the windbreak, which reduces its efficiency).

Gaps in windbreak hedges

If you have a gap in a windbreak hedge (e.g. your main access on to the site), then wind speeds up by about 15 per cent as it goes through the gap, and anything the wind hits on the inside of the area will get battered. Try to avoid such gaps, if possible, by using access from a less windy direction. Sometimes it is unavoidable, but you can still do something to help plants, such as making a staggered entrance (Figure 6) or planting a short 'baffle' or 'island' windbreak inside your site area (Figure 7).

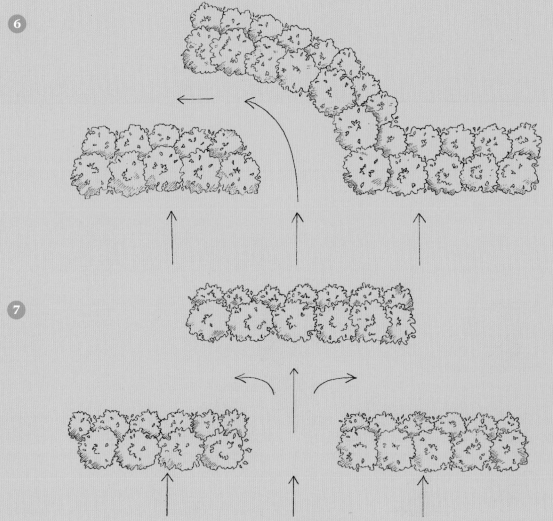

Avoiding wind damage through gaps in windbreaks. **Figure 6** Using a curved entrance way to deflect the wind. **Figure 7** Using a baffle windbreak within the site to absorb the wind.

Windbreaks on slopes

On slopes, avoid making windbreak hedges on contour lines, as you will be creating extra frost pockets where cold air gets trapped and causes colder and more damaging frosts (see 'Aspect' above). Instead plant a hedge just off the contour, i.e. with a slight upward or downward slope as you look along the line of the hedge (see Figure 8). Cold air hitting the hedge will flow down the slope, and if you've left a gap at the lower end it will flow out of the area and further down the slope.

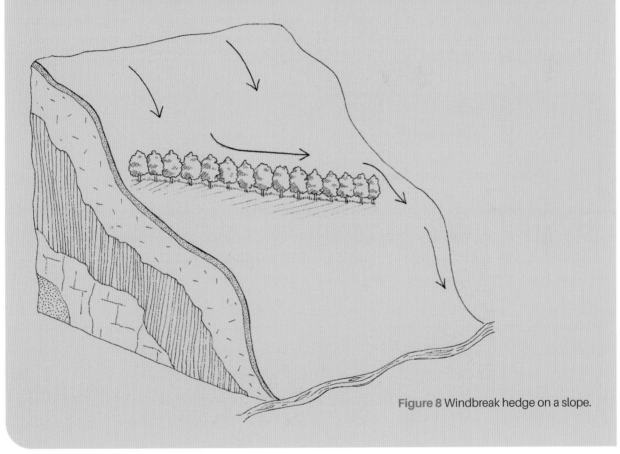

Figure 8 Windbreak hedge on a slope.

Soils

If you don't know it already, get to know your soil. Soil texture (the amount of sand, silt and clay) and soil pH (the acidity or alkalinity) are important factors in determining what you can grow. pH is determined by the availability of calcium and the scale runs from 0 (most acidic) to 14 (most alkaline), with 7 being neutral. Soil test kits are readily available to do your own tests, or you can send off soil samples for a soil analysis. Both are worth doing.

There are some nut trees which can be grown in all types of soil, for example bladdernut or hazelnut. The tables on this and the facing page indicate the types that tolerate different soil textures (light, medium and heavy soils) and those that tolerate different soil pH (acidity/alkalinity) levels.

Almond is the only nut tree that is grown on a variety of rootstocks (see 'Buying and planting trees', page 40) and the main ones are listed in the tables.

Nut tree tolerance of soil types

Light (sandy)	Medium (loam)	Heavy (clay)
Almond (almond/ GF677 rootstock)	Almond (Pixy / St Julien / Myra / GF677 / Marianna rootstock)	Almond (Pixy / St Julien / Myra / Marianna rootstock)
	Black walnut	Black walnut
Bladdernut	Bladdernut	Bladdernut
	Buartnut	Buartnut
	Butternut	Butternut
Chinkapin	Chinkapin	
Ginkgo	Ginkgo	Ginkgo (if well drained)
Golden chinkapin	Golden chinkapin	Golden chinkapin (if well drained)
Hazelnut	Hazelnut	Hazelnut (if well drained)
	Heartnut	Heartnut
	Hickory	Hickory
Monkey puzzle	Monkey puzzle	Monkey puzzle
Oak	Oak	Oak
	Pecan	Pecan
Pine	Pine	Pine (if well drained)
Sweet chestnut	Sweet chestnut	
Trazel	Trazel	Trazel (if well drained)
	Walnut	Walnut
Yellowhorn	Yellowhorn	Yellowhorn (if well drained)

Nut tree tolerance of soil pH

Acid (pH under 6.0)	Medium (pH between 6.0 and 7.0)	Alkaline (pH over 7.0)
	Almond (Pixy / St Julien / Myran / GF677 rootstock)	Almond (Pixy / St Julien / Myran / GF677 rootstock)
	Black walnut	Black walnut
Bladdernut	Bladdernut	Bladdernut
	Buartnut	Buartnut
	Butternut	Butternut
Chinkapin	Chinkapin	
Ginkgo	Ginkgo	Ginkgo
Golden chinkapin	Golden chinkapin	
Hazelnut	Hazelnut	Hazelnut
	Heartnut	Heartnut
	Hickory	Hickory
Monkey puzzle	Monkey puzzle	
Oak	Oak	Oak
	Pecan	Pecan
Pine	Pine	Pine
Sweet chestnut	Sweet chestnut	
Trazel	Trazel	Trazel
	Walnut	Walnut
Yellowhorn	Yellowhorn	Yellowhorn

Most nut trees, just like most fruit trees, crop better in a soil of good fertility (see Chapter 2, page 46). Several tolerate poor soils (including almond, hazel and pine), but their cropping is likely to be reduced.

Beneficial fungi

All nut trees form symbiotic associations with species of mycorrhizal fungi – beneficial fungi growing in association with plant roots. Not only do these fungi significantly aid mineral nutrition (especially of nitrogen and phosphorus), but they also protect their companion plant's roots from attack by soil diseases and pests. Trees growing with such fungi are additionally protected against stress caused by drought and other extreme conditions. In return, the fungi receive some 20 per cent of all sugars photosynthesised by the trees. However, note that soluble chemical fertilizers and soil cultivation have negative effects on these fungi. In time these fungi usually arrive by themselves via airborne spores, but this can take years,

Mycrorrhizal fungi in a chestnut forest.

depending on the site and situation. Sometimes trees bought from tree nurseries will be inoculated with mycorrhizae simply by having been grown in soil where the right types of fungi are prevalent. But there are also methods of improving the chances of having suitable fungi growing with the trees.

A mixture of fungal spores of suitable species can be used in solution as a root dip or as an amendment to planting holes. You can buy commercial spore mixes (see Resources) or you can make your own by harvesting fresh mycorr-rhizal mushrooms from beneath trees – ideally of the same sort as you are planting – and liquid-izing them to make a root dip.

Many mycorrhizal species don't produce edible mushrooms, though there are a number that do. If you can use spores from these as planting amendments, it is likely that you will also get a crop of edible mushrooms beneath your nut trees, which is a great bonus crop. There are now commercial spore mixes available using all edible species of fungi. Of the non-edible fungal species, *Glomus* spp. associate with many types of tree.

It is possible to purchase hazel and oak trees inoculated with species of edible mycorrhizal fungi, including truffles and others. The trees are expensive and come with a 'guarantee' of successful inoculation. However, this doesn't

Edible mycorrhizal fungi associated with nut trees

Chestnuts:

Boletus edulis - cep, penny bun

Gyroporus castaneus - chestnut bolete

Hygrophorus marzuolus - March woodwax

Russula aurora - dawn brittlegill

Tuber spp. - truffles

Hazelnuts:

Tuber spp. - truffles

Uloporus lividus

Oaks:

Amanita caesarea - Caesar's mushroom

Boletus aereus (bronze bolete, dark penny bun), *B. aestivalis* (summer king bolete), *B. appendiculatus* (butter bolete), *B. pulverulentus* (inkstain bolete), *B. regius* (king bolete)

Cortinarius praestans* - Goliath webcap

Exsudoporus frostii* - Frost's bolete, apple bolete

Gyroporus castaneus - chestnut bolete

Hygrophorus marzuolus - March woodwax

Leccinum crocipodium - saffron bolete

Leccinum quercinum - orange oak bolete

Russula brunneoviolacea, R. xerampelina (crab brittlegill)

Sparassis laminosa

Pines:

Boletus pinophilus - pine bolete

Hygrophorus camarophyllus - arched woodwax

Lactarius deliciosus - saffron milkcap

Lactarius deterrimus - false saffron milkcap

*These species are easily confused with poisonous species.

guarantee you will ever get edible mushrooms! There has been a lot of hype about growing truffles, but there are also an increasing number of reports of failure to get any. Truffle trees usually need a well-drained alkaline soil, so with acid soils you can practically rule them out. One cause of failure of pre-inoculated trees to lead to edible mushrooms is that other localized mycorrhizal species outcompete the 'main' fungal species you want, which can sometimes vanish completely.

Some species of mycorrhizal fungi with good edible mushrooms are known to associate especially well with specific nut trees, and these are listed in the table below. **Always be sure you know what you are eating when harvesting mushrooms from beneath trees.**

Pollination

In the context of nuts, planting a single tree is uncommon because only a few species are self-fertile (see table on page 30). Most nut trees produce both male and female flowers (i.e. are *monoecious*), but the timings don't overlap, so other trees are need for cross-pollination. If male flowers open before the females, the plant is described as *protandrous*; if female flowers open before the males it is called *protogynous*. Occasionally a species is *dioecious*, i.e. male and female flowers are borne on different plants, so both sexes of plant are needed for pollination and cropping to occur – monkey puzzle comes into this category.

Apart from those listed in the table, a minimum of two different plants must be grown for most nut species. Sometimes these must also be selected for flowering compatibility – for

Sweet chestnut flowers

Self-fertile nut trees

Almonds (some cultivars)
Bladdernut
Sweet chestnut (few cultivars)
Walnut (some cultivars)
Yellowhorn

example a protandrous cultivar with a protogynous cultivar. See also Chapter 2, page 60.

Sometimes one cultivar is grown as a particularly good pollinator, as is often the case with sweet chestnuts. If you are using one tree as a good pollinator for others and the tree is wind-pollinated, try to place it on the prevailing wind side of the other trees to ensure pollen is blown (gently!) to the trees that need it.

On a larger scale, a nut orchard can be of a single species or several mixed together, but make sure the planting arrangement (see opposite) allows for good pollination. Each tree should have a pollinator tree within three trees' distance away. Mixing species together can be problematical if they all grow to different sizes because in due course the larger ones will start shading the smaller ones.

Planning your site

On a small scale, planting arrangements can be very flexible. You are only likely to have a few nut trees, and pollination, whether by insects or the wind, is not a problem over a short distance. The main factor to consider is availability of sunlight – make sure that in time any larger nut trees are not shading out smaller trees that you want to retain.

A number of species – for example, sweet chestnut and the whole walnut family – grow into large trees eventually and the correct planting distances for each final tree are large – up to 15m (50') or so. Trees planted at this spacing appear to the human eye much too widely spaced and it is easy to think 'Oh, I'll squeeze them up a bit' – this can be a big mistake! Make sure you keep final tree spacings large enough. This does not stop you from using the space in between those trees for something else, though, whether it is fruit or nut trees that will later be removed, or some other cultivated crop. See below for details of the options you have.

Commercial non-organic nut orchards often have the ground herbicided below trees to maintain bare ground or to create bare ground by the time nuts are harvested, making harvesting easier. However, on a garden scale and in any sustainable type of nut growing, there will be something growing beneath the nut trees and this needs planning.

The simplest arrangement is to have grasses beneath nut trees. These can be grazed or cut from time to time, and especially cut low just before nut harvest.

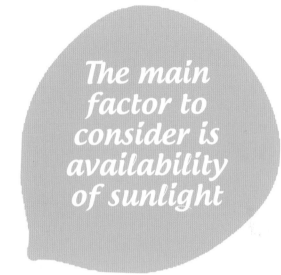

The main factor to consider is availability of sunlight

Planting arrangements

Planting arrangements on an orchard scale can vary – most commonly a square arrangement (see Figure 9) or triangular arrangement (see Figure 10). If the tree spacings are the same, the triangular arrangement fits a few more trees into the same area than the square arrangement.

For many nut trees, the final planting distances are generous to allow for the large trees to fill the space. For example, with sweet chestnut and walnut the final tree spacing might be 12-15m (40-50'). These spacings leave a lot of empty ground between trees for several years, and this is an opportunity to grow something else.

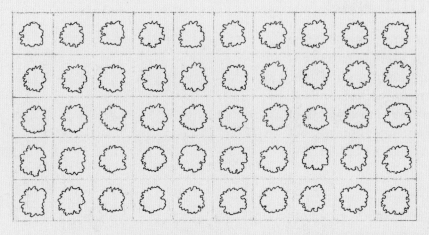

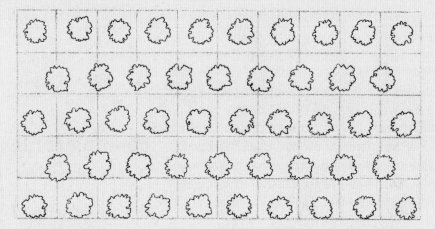

Figure 9 Square planting arrangement. Diagonal distances are slightly greater.

Figure 10 Triangular planting arrangement. Trees are equidistant from each other.

Alley cropping

There are several traditional interplanted systems that have long been used in nut orchard planting. These include growing hazelnuts with alleys of grape vines or fruit bushes between, and growing alleys of annual vegetables between hazel, sweet chestnut or walnut trees. There can be both ecological and economic benefits of interplanting: physical resources are used efficiently, and shorter-term crops can provide income while the longer-term crops mature.

One option, used increasingly on large-scale plantings, is alley cropping of nut trees (see Figure 11). This involves planting the trees at final spacings in widely spaced lines in a field. In between the lines, alleys of annual crops are grown – cereals, annual vegetables, etc. The width of the alleys is chosen as a multiple of the width of the equipment used. A minimum of 2m (6'6") width is uncultivated along the tree rows (the ground cover / weed growth in uncultivated areas can be a management

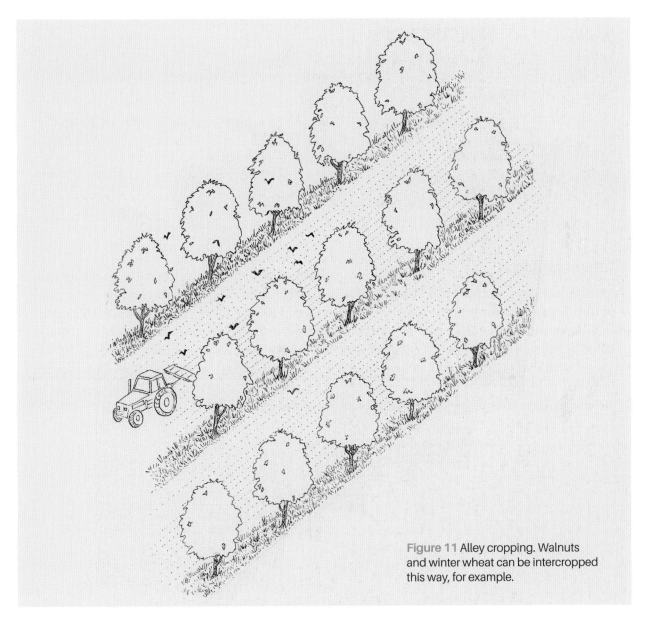

Figure 11 Alley cropping. Walnuts and winter wheat can be intercropped this way, for example.

issue). In time, as the trees grow, production from the edges of the alley reduces, mainly due to shading, and in small alleys under 15m (50') wide, a rough rule-of-thumb is that arable production remains good for the same number of years that the alleys are wide in metres: so with alleys 12-15m (40-50') wide,

shading will make arable intercropping unviable after 12-15 years. Alleys can be quite large (e.g. 30m/100' or more) to maintain alley crop yields. The crop grown in the alley is chosen so that it is harvested by the nut cropping time, allowing for access beneath the trees to harvest the nuts.

Nitrogen-fixing nurse crops

Alder	Shrubby - *Alnus sinuata* (Sitka alder), *A. viridis* (green alder) Tree - *Alnus cordata* (Italian alder), *A. glutinosa* (alder), *A. rubra* (red alder – very fast-growing)
Broom	*Cytisus scoparius*
Elaeagnus	*E. angustifolia* (oleaster), *E. commutata* (silverberry), *E. multiflora* (goumi), *E. umbellata* (autumn olive)
Sea buckthorn	*Hippophae rhamnoides*, *H. salicifolia* (Himalayan sea buckthorn)
Siberian pea	*Caragana arborescens*
Wax myrtle	*Myrica californica* (Californian), *M. cerifera*, *M. pennsylvanica* (northern)

Intercropping

Another option is to interplant another tree or shrub crop in between the nut trees. The interplant could be a fruit tree or shrub, or could be the same type of nut tree that is being grown at full spacing, which is to be removed (thinned out) later. It could also be a nitrogen-fixing nurse tree or shrub, which will boost the growth of the nut trees (see table above). Although this system requires more capital input for the extra trees, there is also a faster return from the intercrop, which should quickly pay for the intercrop trees. An example is shown in Figure 12 where a square planting arrangement of nut trees has rows of an intercrop.

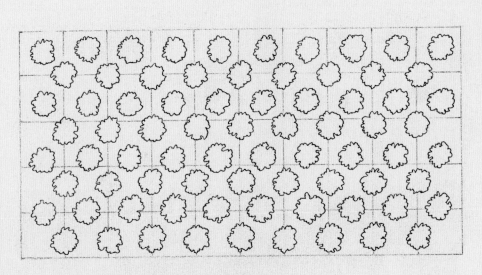

Figure 12 Square arrangement planting with rows of an intercrop.

Sweet chestnut burrs. Different cultivars often have distinctive burrs.

Eventually, of course, the intercrop will be shaded out by the nut trees and the intercrop trees will die or need to be cut out and removed. One of the main questions for this system is how long will the intercrop trees crop for before they are outshaded? It depends on the nut crop, spacing, your soil and other factors, but some likely figures are given below.

Also be aware that the walnut and hickory family have the chemical juglone in their leaves and roots, which can suppress plants – noticeably apples, some pines, and the potato and legume families. Plants are normally only affected when the tree canopy is actually over-head – when the chemical can kill off those plants.

Intercrop productivity periods

Nut crop	Nut tree spacing (square)	Productive life of intercrop trees
Black walnut	12-15m (40-50')	12-25 years
Buartnut		
Oak		
Pine nut		
Walnut		
Heartnut	12-15m (40-50')	10-15 years
Sweet chestnut		

A slightly more intensive intercropping arrangement plants in rows between the main nut trees but also the gaps between nut trees (see Figures 13 and 14). This requires more intercrop trees but uses the space more efficiently and productively.

Interplanting need not inevitably lead to a practical monoculture of nut trees. Trees can be underplanted with some shade-tolerant crops that don't die out over time, and trees can also be planted at more than the minimum final planting distances to allow for interplants to continue to get good light. In fact, diversity is

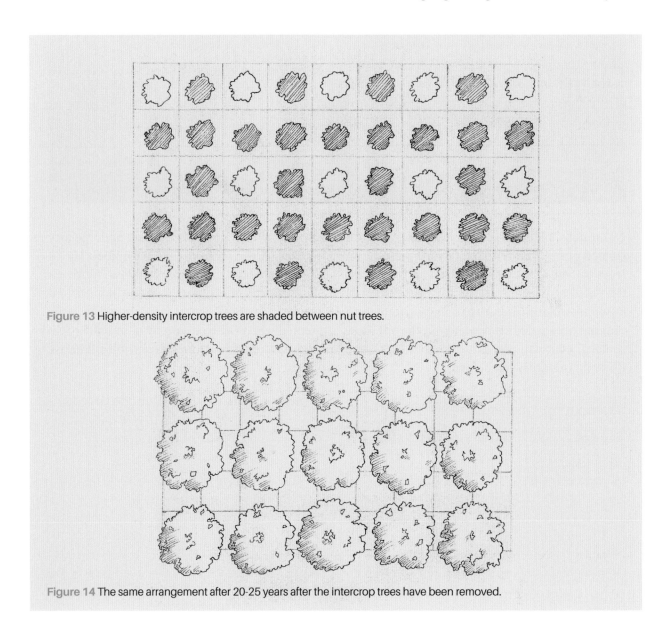

Figure 13 Higher-density intercrop trees are shaded between nut trees.

Figure 14 The same arrangement after 20-25 years after the intercrop trees have been removed.

one key to sustainability so this is good practice, but not always easy to achieve. Nuts are usually harvested from the ground; this means that any low perennial intercrop needs to be either cut before harvest or covered with a net at that time. The last thing you want to be doing is searching among low plants to find nuts. Figure 15 shows an example, where forest-garden-type perennial underplanting is used beneath nut trees. Some perennials that tolerate shady conditions beneath trees include edibles – e.g. wild garlic (*Allium ursinum*), Siberian purslane (*Claytonia sibirica*), ground-cover raspberries (*Rubus* spp.); and medicinals – e.g. lady's mantle (*Alchemilla mollis*), black cohosh (*Cimicifuga racemosa*), ginseng (*Panax ginseng*), lungwort (*Pulmonaria officinalis*), comfreys (*Symphytum* spp.), periwinkles (*Vinca* spp.).

Figure 16 shows another permanent underplanting scheme, using smaller fruit trees between the larger nut trees that are planted at a wide spacing. This example shows grass beneath all the trees and this would be cut or grazed, although planting a perennial ground layer is feasible to make a kind of 'nut forest garden'. Grasses beneath nut trees can, of course, be cut to make silage, haylage or hay.

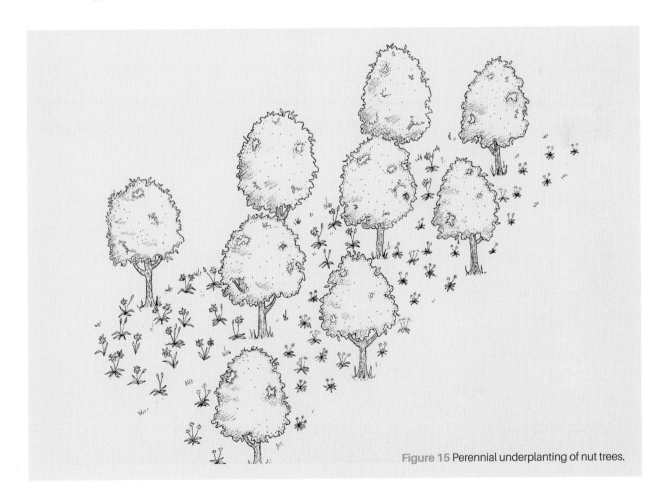

Figure 15 Perennial underplanting of nut trees.

Figure 16 Fruit and nut tree permanent planting scheme. Lower plan view indicates nut trees with shading.

Silvopasture

Planting nut trees in pasture ('silvopasture') is a traditional agroforestry system that still continues to be used in Europe. Like other intercropping systems, there can be ecological benefits (e.g. soil protection), economic benefits (animals to sell before nuts mature) and also benefits to the animals (shade, shelter, etc.). Sheep are the main animal used in undergrazing systems, as with good management they cause little damage to the trees or soil (see Figure 17). Cattle and trees do not mix so well – at least until trees are several decades old. Geese are another option.

If you intend to use sheep beneath trees, then firstly consider the breed. Breeds differ in their gregariousness (clustering) instinct. Black-faced (meat) breeds tend to be preferred because they are less gregarious and their lambs have more rapid growth than white-faced (wool) breeds. Undesirable behavioural habits, for example browsing, tend to spread rapidly through gregarious flocks. Breeds that have been used successfully include Dorset, Hampshire, Leicester, Rambouillet, Romney, Shropshire and Suffolk.

Sheep flocks are most difficult to manage in the spring, when the growth of cool-season grasses accelerates and trees are highly susceptible to browse damage. Forage availability and browse damage should be monitored daily, and frequent flock rotation – perhaps every three days

Figure 17 Grazing sheep beneath nut trees.

or so – may be necessary. Sheep may, through boredom, cause more damage the longer they remain in one pasture. They will also cause more damage as forage becomes limited.

Tree browsing is a learned behaviour, so it is important to get rid of these 'lawbreakers' before they teach others their miscreant ways. Almost always, a relatively small number of animals in the flock cause all the damage and when removed, damage ceases. This advice also applies for 'jumpers' that ignore fences.

Avoid putting weaned lambs in with trees by themselves, as lambs challenge and butt into trees, and investigate by browsing more than ewes. The frequency of flock rotation should be adjusted to minimize tree damage. Lambs by themselves take much more time and management for use in silvopastures; however, they tend to follow the example of older sheep in a pasturing routine. Dry ewes are best for this system, but ewe–lamb pairs may be used if necessary.

Sheep will seek shade during the heat of warm summer days and if none is provided, they will lie under larger trees, possibly damaging lower branches. Ideally, some shade near the water supply would be provided, and the sheep will naturally rest there after watering.

It is also important to control external parasites, as they can stimulate a rubbing response.

To optimize tree and sheep management, clover mixtures tend to be the best forage – for example, white clover with perennial ryegrass, etc.

Using grazing animals beneath nut trees will not leave the orchard floor dead flat – meaning that larger-scale harvesting equipment will function less efficiently.

Buying and planting trees

Although nut trees can be propagated, most of us will be purchasing the nut trees we want to grow.

Types of tree to plant

Some nut trees are usually planted as seedlings, particularly if there are no improved fruiting cultivars – yellowhorn, for example. So for these you'll need to buy seedling plants or raise them yourself, usually from seed. For nut trees that have had cultivars selected, you'll usually want to use some of these in your planting to ensure known good nut production. A few of these, like hazels, are often propagated on their own roots, but most will be grafted trees, using a closely related or same species rootstock.

Although many nut trees are propagated by grafting on to rootstocks, like most fruit trees, there has been very little work gone into selecting rootstocks for nut trees. Hence most that are grafted – for example chestnuts and walnuts – are usually grafted on to ordinary seedlings of the same species. Usually only almond is grafted on to a variety of rootstocks and this is because the rootstocks developed for peaches and plums are mostly compatible with almond too. See Chapter 2 for more on nut tree propagation.

Buying

When buying nut trees, bear in mind that most nurseries supplying nut trees are specialist small-scale nurseries with limited stocks, so order early to ensure you get what you want. Bare-rooted trees (grown in a field and dug

A multi-stemmed walnut tree.

when dormant) or pot-grown trees can be used. Most nut trees are sold as one- or two-year-old maidens (trees with a single straight stem with few or no side branches for 1.5-3m/5-10') and will be dispatched when dormant over the winter. The best time to plant the trees depends on your climate: in mild temperate climates, autumn planting is good, although any time in winter is fine; in colder winter climates, spring planting is the norm. Spring-planted trees are more likely to need irrigation in the first couple of summers, should they turn dry. Although larger trees (especially walnuts) can sometimes be purchased, it isn't worth it, as they are a lot more expensive and in the long run gain you nothing: smaller trees catch up quickly. Larger trees are more difficult to stake as well.

Planting

Mark out the planting positions with sticks or canes according to your planting plan – don't guess distances but measure to ensure proper spacing.

Dig a planting hole wide and deep enough so the roots are not crammed into it. Try to loosen the sides of a planting hole to make it easier for the roots to grow horizontally. Once the hole is dug, a stake can then be hammered in – most nut trees, like fruit trees, will benefit from staking in their early years. If the tree is pot-grown and it has become a bit pot-bound with circling roots, cut through these with a knife to force new growth outwards from the root ball. Try to attach the tree quite low to the stake

Planting and mulching a nut tree

1. The planting hole dug in a square shape – this is usually better than a round hole, as the roots do not circle.

2. The tree placed in the hole, to make sure it is the correct depth.

3. The hole backfilled with earth.

4. Thick cardboard is fitted around the tree to kill off surrounding grass. Bark mulch is laid on top of the cardboard to (a) hold it down, (b) provide extra mulch and (c) improve aesthetic appearance.

(30-60cm/1-2'). The example in the photos (opposite) is a small hazel tree, which does not require staking. Always mulch newly planted trees – here the hazel tree has been mulched with card, to kill off the surrounding grass, and the card covered with bark mulch.

If the soil is of reasonable fertility, there's no need to add any kind of amendment, apart from mycorrhizal fungi if you are using them, so just backfill with the soil you took out.

You may need to protect trees from browsing animals – rabbits, deer, etc. Some species are more susceptible than others to browsers eating off leaves and shoots (chinkapin, hazel, oak, sweet chestnut, trazel and yellowhorn are all more susceptible) but a hungry rabbit might eat anything! Ordinary spiral or tube tree guards are fine against rabbits. Deer need higher guards (1.2m/4' for roe deer, 2.1m/7' for red deer), which are trickier to use, especially if there are low branches. You may need to make your own larger guards out of netguard plastic mesh or wire mesh, using two or three stakes.

You'll need to protect trees from deliberate browsers like sheep or cattle, which are to undergraze a nut orchard. One (expensive) option is to create a fenced enclosure 1-3m (3-10') across using 3-4 posts and wire or wire netting. Alternatively, narrow net or wire guards are fine (1.5m/5' for sheep, 2.1m/7' for cattle), but the posts need to be stronger – enough that sheep cannot push them over if they get their front feet up on them.

> *Dig a planting hole wide and deep enough so the roots are not crammed into it.*

Nut trees planted on public urban sites should be staked and protected well, as vandals will wreck them just as they will other trees on a whim.

Dig a planting hole wide and deep enough so the roots are not crammed into it. Try to loosen the sides of a planting hole to make it easier for the roots to grow horizontally. Once the hole is dug, a stake can then be hammered in – most nut trees, like fruit trees, will benefit from staking in their early years. If the tree is pot-grown and it has become a bit pot-bound with circling roots, cut through these with a knife to force new growth outwards from the root ball. Try to attach the tree quite low to the stake (30-60cm/1-2').

Maintenance & propagation of nut trees

If you're the sort of grower that likes to plant a tree and let it do its own thing, don't worry – you can do that with many nut trees and they will be fine. But if you want to maximize your crops, have regular crops and keep the nut quality high, then a certain amount of maintenance is recommended. Some feeding once trees are cropping well is a good idea, and if you have dry summers, then irrigation will make sure the nuts grow to a normal size. Pruning – especially in the early years – is sensible, and there are pests and diseases that can be avoided or managed to avoid serious crop losses. This chapter covers all these maintenance aspects before looking at nut tree propagation. Cultivating your own plants isn't always easy, but there are ways to overcome the challenges.

Feeding

Feeding your nut trees on a small growing scale is normally a very different proposition than that facing a commercial grower with a monoculture orchard. In the latter there is often no input from any other plants and the nut trees are heavily fed, leading to both heavy nut production and heavy infestations of pests (requiring more interventions). Mainstream agriculture, including monocultures of nut trees, wastes a huge amount of the fertilizer used – a large proportion of which is usually washed right out of the soil into watercourses.

On a smaller / more sustainable scale, a general rule is: If the trees are cropping well without feeding – don't feed. Only feed if the nuts are smaller than usual, the crop is reduced, or the trees significantly slow their growth. If trees are growing with associated mycorrhizal fungi (see Chapter 1, page 27), these will improve tree nutrition tremendously.

Like most flowering and fruiting plants, nut trees require a good supply of potassium

Soil pH also has a significant effect on nutrient availability: if the soil gets too acid or too alkaline, then any nutrients become locked up and plants cannot access them. It's important to keep an eye on that and if appropriate amend with lime. Adding a liming material (e.g. ground limestone) increases the soil pH. Although agricultural soils tend to acidify over time in a wet climate, deciduous tree crops counteract this tendency, so as trees establish, the soil pH should stabilize in these situations. If you decide to lime, note the requirements listed in the table in Chapter 1, page 27 and use soil analysis results to work out how much lime you need to apply. Materials like ground limestone can be broadcast straight over the ground.

Like most flowering and fruiting plants, nut trees require a good supply of potassium and nitrogen in particular.

The most sustainable way of providing nitrogen is via nitrogen-fixing plants (see Chapter 1, page 20). If you can use these – as windbreaks, as interplants or as part of the ground-cover layer – then you are well on the way to being self-sufficient in nitrogen. You might have other sources such as compost, manures, etc., which should be spread in spring beneath the trees; using cut grass mulches also supplies lots of nitrogen. Urine is a fine fertilizer, containing about 11g of nitrogen per litre, and depositing it under nut trees is an ideal way of using it! (1 litre / 1¾ pints) of urine will sustain about 200-250g / 7-9oz of nut crop; 1kg / 2lb 2oz of compost or manure will sustain about 100g/4oz of crop.)

Potassium is needed throughout the growing season and can be applied from spring to early autumn. Applications are fairly quickly stabi-

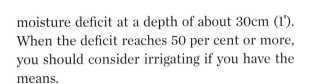

lized in the top soil layers. If you use urine, manures or composts, these will all contain significant amounts of potassium. Wood ash is another good source, containing 5-15 per cent potassium, but should be used in several light applications rather than a heavy one, as the potassium content is very soluble and easily washed out of the soil.

Some potassium can also come from potassium accumulator plants. These plants are very efficient at extracting potassium from the soil or subsoil, and make it available to other plants in the topsoil layers when leaves die back or plants are mown/cut, etc. Comfrey is one of the best known and can be cut low before harvest if grown beneath trees. Other potassium accumulators that can tolerate mown conditions beneath trees include plantains (*Plantago* spp.), buttercups (*Ranunculus* spp.), docks and sorrels (*Rumex* spp.) and dandelions (*Taraxacum officinale*).

Irrigation

Although irrigation is often not needed in cool humid climates, it can sometimes be useful (especially for young trees); and in warmer, drier climates it can be essential.

Irrigation is likely to be most important on sandy and sandy loam soils, and the most critical time to water nut trees is mid- and late autumn when nuts are swelling. If these months (and the ones preceding it especially) are dry, then there will be a moisture deficit in the soil, which will cause nuts to remain small.

To know if irrigation is necessary to avoid loss of crop, you need to measure or judge the soil moisture deficit at a depth of about 30cm (1'). When the deficit reaches 50 per cent or more, you should consider irrigating if you have the means.

There are various ways to find the soil moisture deficit:

* **Soil appearance.** The simplest way is via the look and feel of the soil. Take a sample of soil from 30cm (1') deep and do 2 tests: form a ball by squeezing the soil hard in your fist; and form a ribbon by rolling the soil between your thumb and forefinger. Depending on your soil texture, then if the following conditions apply, the deficit is around 50 per cent:
  - **Sandy / loamy sand:** appears dry; forms very weak ball or will not ball.
  - **Sandy loam / fine sandy loam:** slightly dark; forms weak ball.
  - **Silty loam / clay loam / silty clay / clay:** balls easily; small clods flatten rather than crumble; ribbons slightly.
  - **Other loams:** fairly dark; will form ball; slightly crumbly.
* **Weighing and drying.** This is a slow but accurate method. Sample the soil, measure

the volume, weigh, dry and weigh again. The soil should be dried in the oven at just over 100°C (212°F) for 24 hours.

Volume % of water = BD x 100 x (wet weight-dry weight) / dry weight, where BD = bulk density of soil: about 1.4 for sandy soils, 1.6 for clay soils.

Use the table below as a guide. If the volume % in the table is less or equal to the above figure for your soil, then irrigate.

* **Soil probe.** Accurate but quite expensive, a soil probe gives a measure of the volume % of water. Use the table below. If the volume % in the table is less or equal to the above figure for your soil, then irrigate.
* **Tensiometer.** Accurate and less expensive, a tensiometer gives a reading of soil tension – a measure of the energy needed for plant roots to extract water from the soil. Use the table below; if soil suction in the table is greater or equal to the tensiometer figure for your soil, then irrigate.

Soil type	% water volume	Soil suction
Clay	35	215 centibars
Silty clay	32	185 centibars
Silty clay loam	29	145 centibars
Clay loam	27	110 centibars
Sandy clay loam	25	82 centibars
Sandy silt loam	24	68 centibars
Silt loam	22	55 centibars
Sandy loam	19	40 centibars
Loamy sand	16	25 centibars
Sand	10	15 centibars

Dripper on a drip irrigation system.

A 50 per cent soil moisture deficit is equivalent to the approximate soil water volume percentage and soil suction shown in the table.

If you do decide to irrigate, then if possible apply water by trickle or drip irrigation to avoid wetting foliage and economize water usage. Low-pressure trickle systems are available if you are using stored water from water butts or small reservoirs, whereas dripper systems usually need water at mains pressure.

Management under nut trees

Most nut trees are deciduous, so in early spring there is plenty of light to enable vigorous growth of grasses or understorey plants. On a garden scale, you might have early crops there, such as wild garlic (*Allium ursinum*), that allow you to harvest other things beneath your trees at this time of year.

If you are grazing beneath trees, then spring up until midsummer is the time when forage is at its greatest, although in mild winter climates, some winter forage should be available too.

If you are mowing regularly beneath trees, spring is when most frequent mowing is required. Mowings can be allowed to drop where they are cut, or to concentrate fertility they can be collected and used to mulch beneath trees.

From midsummer to autumn, trees will be in full leaf and casting significant shade. This will reduce the growth of anything underneath – how much depends on tree spacing (see Chapter 1, page 31).

Any feeding of the understorey should take place mostly through the spring. Fertilizers/manures will of course also feed the nut trees (see 'Feeding', page 46), and in a cool autumn climate you don't want to encourage late growth of trees that will struggle to harden off (fully ripen and become hardy) before winter, because such growth is very prone to dieback – when new shoots die back over winter, becoming obvious when growth starts in spring.

Nut harvest is mostly mid- and late autumn (depending on the species) and is from the ground, apart from the smaller shrubs and immature hazelnuts. You might have a variety of things growing between and beneath your nut trees, of course, so whatever is growing there should not interfere with harvest. In practice this means that ground cover needs cutting low or that nets are used to harvest (see Chapter 3, page 66).

Pruning

As for most fruiting trees, in the first few years it is worth doing a little formation pruning – initital pruning to select the main branches – but after that nut trees require little pruning on the whole. The majority of nut trees are tip-bearers (i.e. producing flowers and fruit on young wood around the canopy surface of the tree) so annual pruning of young wood is NOT what you should be doing, as this will remove the flowering wood and future crop.

The general rules for formation pruning are described below and shown in Figure 18.

Prune off very low branches – these will interfere with harvesting and access underneath trees. Work upwards over 3-4 years, going no higher than halfway up the tree, until you get to a desired height for the first branches.

Prune off very upward-growing branches and those with an acute (small) angle between the main trunk and the branch. These branches will rub against other branches, causing damage, and will be susceptible to breaking in strong winds when they are larger.

Sweet chestnuts are more vigorous in growth than most of the other nut trees and may need a little more formation pruning. Figure 19 shows the typical pruning needed for a new chestnut tree in the first three years.

For sweet chestnuts (and any other vigorous-growing fruiting tree with a tendency to upward growth), it may even be worth tying down some of the lower branches, as shown in Figure 20. The branches are tied to stakes for a year or two, using rubber ties around the branches so as not

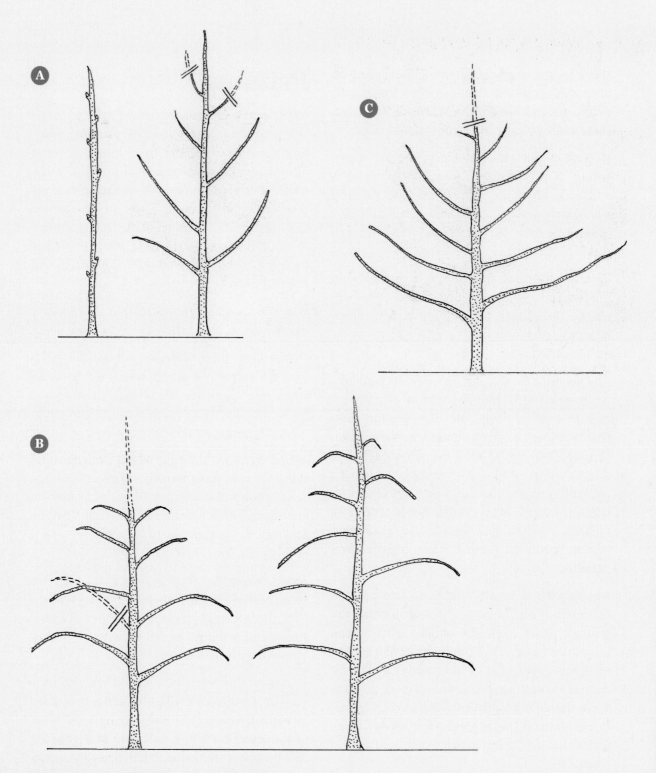

Figure 18 Formation pruning – general principles.

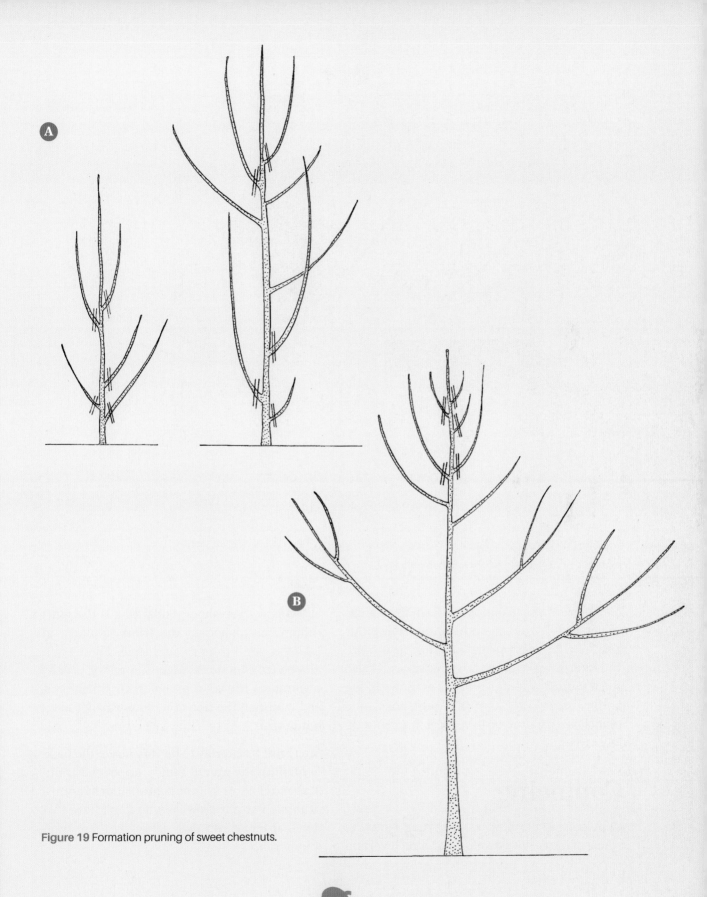

Figure 19 Formation pruning of sweet chestnuts.

Figure 20 Tying down sweet chestnut branches.

to damage them, to pull them down to an angle of about 30° from the horizontal. Doing this keeps the lower branches from turning upward and interfering with the branches above. It also speeds up flowering/fruiting, as flowers are borne more quickly on near-horizontal branches than on near-vertical branches in most trees.

Coppicing

Coppicing is the process of cutting back the whole of a tree low down near ground level and allowing it to re-shoot. Pollarding is the same thing done higher up the trunk, typically at 1.2-1.8m (4-6') from ground level. These techniques are commonly used in forestry, usually to produce firewood or poles for other uses, and maintain the tree at a much smaller maximum size.

Many nut trees grow to be very large (the hickory family, the walnut family, sweet chestnuts, etc.) and cannot easily be kept smaller by annual pruning. For nut growers who don't have that much space, it can make growing these trees impossible.

One option for keeping these trees smaller is to coppice or pollard them on a regular basis, for example every 8-10 years. There are advantages and disadvantages to this:

Advantages
* Virtually no annual pruning is required.
* Trees are kept shrubby and smaller.
* Branches are easily shaken to release nuts.
* Any biennial cropping habit is lessened.
* Coppiced trees can live to a great age.
* Source of firewood.

Disadvantages
* Initial growth after coppicing is quite upright.
* No cropping for a few years after each coppicing, although heavier cropping later in the cycle largely compensates.
* Grafted trees may sucker (send up new shoots from below ground level) or send up rootstock shoots which need controlling.
* Branches may be more susceptible to wind damage, as many will be upright.
* Large pruning cuts at coppicing might increase the risk of diseases.

A walnut coppiced every 10 years will reach a height and width of about 5-6m (16-20'), and a sweet chestnut will reach 6-8m (20-26').

Smaller nut trees like hazelnuts can also be coppiced in a cycle to keep trees smaller and to reinvigorate older trees that have become unproductive.

Main pests

Specific pests and diseases are described for each nut tree species in Part II. However, there are a few pests that affect all nuts in many parts of the world – namely squirrels and birds.

Squirrels

In some areas of Europe and North America, grey squirrels (*Sciurus carolinensis*) are the problem. In much of mainland Europe, red squirrels (*S. vulgaris*) are considered pests. In North America, the native red squirrel (*Tamiasciurus hudsonicus*) is also a problem.

Grey squirrels will not only consume the nuts from your trees, but outside North America they will also damage the trees by bark stripping.

Grey squirrels mainly frequent woods containing broad-leaved trees, and in the winter they live together in large nest-like structures (dreys) made from leafy branches and constructed high in the branches of trees. The female does live alone in large breeding dreys, or in holes in trees, and produces litters of 3-4 young, either in late winter and/or in midsummer.

Red squirrel (*Sciurus vulgaris*) - protected in the UK

Grey squirrel (*Sciurus carolinensis*) with walnut.

Damage takes the form of bark stripping from young trees (normally 5-40 years old). This is concentrated around the base, and at the points where side branches meet the trunk (where the animals can perch), with both trunk and branches stripped. Patches of bark are missing and some strips might still be attached. Mainly young male squirrels strip bark to get to the sap, and damage appears to be triggered by population stress, especially high numbers of males. Risk of damage is greatest when populations are highest, which happens following mast years (when oaks or beech produce large amounts of seed) followed by mild winters and early springs. Most damage occurs between mid-spring and late summer. Susceptible nut trees include oaks, pines and the walnut family. Sweet chestnut is also sometimes attacked.

All squirrels can be serious pests of all nut crops, but most at risk are hazelnuts and walnuts, which are both taken from the trees before they are fully ripe. Hickories, oaks and pines are also readily predated. Chestnuts have spiny burrs that act as a deterrent, so tend to be attacked only when the burrs open and nuts start dropping. If you have a large nut-producing area, damage tends to start at the edges of the area, moving inward gradually. One squirrel can account for 25kg (55lb) of nuts, some eaten immediately and others buried for the future. That's the crop from one mature walnut tree.

If you want to grow nuts, you have to protect them from squirrels, and a number of approaches are listed below.

Coping methods

❋ Harvest daily so that nuts are not sitting on the ground for long.

❋ Shake/knock branches before each harvest to release loose nuts.

❋ Don't site nut trees next to woodland.

Deterrent methods

❋ On a garden scale, fruit cages and netting will deter squirrels; permanent nut cages for nut shrubs are best made out of wire netting, as squirrels easily eat through plastic nets.

❋ Electric netting, 1.5m (5') high with a 10cm (4") mesh, can sometimes keep squirrels out of an area for a while. Occasionally animals might leap through the mesh, and you need to be vigilant about plants short-circuiting the netting.

❋ Use of artificial snakes in trees can be effective for a short period.

❋ Squirrels can jump 3m (10') horizontally from tree to tree, so try to ensure that there is a gap larger than this around the boundary of your nut-growing area.

❋ Place tree guards around trunks. Squirrels can jump 1.5m (5') vertically, so if you want to stop them climbing into trees, first make sure that all branches are higher than this. A guard made of a hard and slippery material – for example, tree-tube plastic – can stop them climbing, but it must be placed so they cannot jump past it, e.g. between 0.6m (2') and 1.8m (6') high around the trunk.

❋ Keep ground cover / grass short.

❋ Allow dogs to run beneath the trees.

❋ Station Grandma with a shotgun beneath the trees!

There is no evidence to suggest that ultrasonic devices will prevent squirrel damage, nor do hawk-like kites. You could use 'sacrificial' trees, for example sycamore (*Acer pseudoplatanus*), to try to tempt squirrels away from your nut trees, but this bears the risk that squirrels will first strip bark on those trees before moving on to your nut trees.

Repellents

There are commercially available squirrel repellents, which are usually based on capsaicin – the compound that gives chilli peppers their heat. You can also sprinkle cayenne pepper underneath trees. Repellents are sprayed on to foliage and the repellent effect lasts only up to two weeks or so – less in rainy weather. If you have only one or two nut shrubs, then it might be worth considering, but on a larger scale and for larger trees, forget it.

Control methods

Some potential nut growers don't like the idea of having to kill squirrels, but the plain fact is that if you have a large squirrel population nearby, you are unlikely to get many nuts without doing so. Note that in certain regions, for example Great Britain, red squirrels are protected and only grey squirrels can be controlled.

❋ **Cats.** Domesticated cats are sometimes good hunters of squirrels. The cat probably needs to be on the large side, as squirrels are extremely aggressive when caught or in fights. Exactly how one would train or encourage a cat to attack squirrels is another question!

* **Dogs.** Jack Russells can be good squirrel hunters.
* **Hunting with hawks.** Two fine predators of grey squirrels are Harris and Redtail hawks. These are native to North America, where the grey squirrel is a normal part of their prey. Some falconers fly these birds and offer a service where they and their birds will come and hunt squirrels.
* **Shooting.** Traditionally, dreys are poked in winter to make squirrels appear so they can be shot. Normally a two-person operation.
* **Cage trapping.** This uses a baited cage to capture an animal alive, which is then killed. If anything else is caught accidentally, it can be released alive. Use from spring to autumn. Single or multi-capture (i.e. can hold more than one animal) cage traps should be set out along the routes squirrels take – typically along hedge or windbreak lines – at a spacing of 50-100m (55-110yds). Bait with yellow whole maize (or the nuts you are harvesting) and cover traps with a plastic sheet

secured with branchwood to encourage animals in and protect captive animals. Cage traps, once set, must be visited regularly and captive squirrels must be humanely killed. As a bonus, squirrels caught this way can be eaten!

Birds

In many countries larger birds can sometimes be a problem by taking ripe nuts. These can include rooks and crows (*Corvus* spp.), parakeets, jays, magpies, starlings and blackbirds.

Damage is caused in autumn as the nuts begin to ripen. Almonds, walnuts and hickory nuts are most at risk, with damage beginning as the husks split to reveal the ripe nuts within. Typical signs of bird damage are the presence of small pieces of broken nutshells appearing on the ground under the branches. A large flock can take the whole crop quite quickly. One large bird such as a rook can eat 20kg (44lb) of nuts.

If you need to defend your nut trees against birds, the 'coping methods' described for squirrels on page 55 also apply to birds. There are also many effective deterrents, but repellents don't work and control is virtually impossible.

Deterrent methods

The following methods can be used against most bird pests. All of them work, but usually only for a short period (5-7 days), so you need to change deterrent a few times during the nut harvest. Birds usually attack the edge of a crop first, gradually moving further and further in, so many scaring devices can work well if placed at the edge of the crop.

Once birds know where there's a guaranteed meal, they will be very difficult to deal with

Carrion crow (*Corvus corone*) eating a walnut.

Be vigilant and start deterring as soon as you detect a problem. Once birds have fed repeatedly in any given spot and know where there's a guaranteed meal to be had all day and every day, they will be very difficult to deal with.

* Birds have excellent eyesight, which some bird scarers exploit by using an intermittent 100 Hz strobe light, with or without an electronic noise scarer (with associated problems).
* A scarecrow moved at least daily – most birds are wary of people.
* Constant movement can be very off-putting to the birds. Hang up items in trees, such as old CDs or empty milk cartons (slit down the four sides of the carton first and slightly squash to open it up so it catches the wind better). Brightly coloured flags flapping in the wind sometimes work. Reflective windmills, which have reflective blades that flash UV light, are another option.
* Attach large 'eye' balloons to poles above treetops. These balloons are about 75cm (30") in diameter and several per acre can be effective. Try to get them a metre (3') above the trees, and use tethers at least 60cm (2') long. Vary the height and position every 3-5 days.

* Bird scare tape – red and silver aluminium-mylar tape (not to be confused with humming line tape, which is generally not so effective). This can be stretched from treetop to treetop in a random pattern. It doesn't need to be densely spread, as long as there is some tape within, say, 20m (70') of any point. Holographic tape has now been developed that is thicker and noisier (makes a metallic noise as it rattles), which might also be helpful.
* There are reports that a monofilament fishing line can be beneficial, as it seems that birds are not quite sure whether the barrier is there or not. For trees, the method recommended is to attach a pole vertically in the centre of the tree and then make a 'tepee' of lines down to the ground from the top. Protruding branches are not a problem, as birds are repelled in a fairly large area around the line.
* Artificial (decoy) snakes or snake lookalikes placed in trees and moved regularly. Use toy rubber ones or pieces of garden hose.
* Artificial (decoy) birds of prey, for example hawks or eagles. These are best placed on top of poles at treetop level, and again moved regularly. There are commercial eagle kites on a tall pole that soar on a tethered line in the wind. A kite alone may also be a good deterrent.
* Intermittent loud noises (only suitable for rural areas, obviously), for example automatic gas guns (gas cannons) at one per 2-3ha (5-8 acres) in the daytime, as used agriculturally. There are also a number of electronic devices that use the birds' own distress calls (or, increasingly, abstract traffic noises, music, screeches and whines) to frighten them off – these can be very loud (up to 120 decibels). Loud noise deterrents are being viewed as increasingly environment-unfriendly and they are likely to send you and your neighbours mad! They are regulated in many areas and complaints of their nuisance value are rocketing. Other methods can be as effective if rotated.
* Shooting with shotguns is effective at scaring birds off. However, what is often forgotten is that shooting consists of two elements of harassment: periodic loud blasts and the presence of humans. Birds can often learn to disperse as soon as humans turn up and then return when they see them leave. There is also the noise/nuisance problem, as above.

Nut tree propagation

Many types of propagation are possible with nut trees, some of which are easy and others very difficult. Cuttings are generally difficult, seed relatively easy and grafting very difficult.

The main propagation methods are listed in the table opposite. In addition, ordinary layering (pegging down a branch to the ground) works for most species. However, this technique is slow (rooting can take a year or more) and requires low horizontal branches. Air layering is less reliable but can sometimes work. This involves wounding a branch and surrounding it with a moist material wrapped in plastic. Roots then sometimes form where wounded.

Nut tree propagation methods

Tree	Softwood cuttings	Hardwood cuttings	Root cuttings	Suckers	Seed	Grafting	Stool layering
Almond					✓*	✓	
Black walnut					✓*	✓	
Bladdernut	✓		✓	✓	✓		
Buartnut					✓*	✓	
Butternut					✓*	✓	
Chinkapin				✓	✓		
Ginkgo	✓				✓*	✓	
					✓	✓	
Hazelnut	✓	✓			✓*		✓
Heartnut					✓*	✓	
Hickory					✓*	✓	
Monkey puzzle					✓		
Oak					✓	✓	
Pecan					✓*	✓	
Pine					✓	✓	
Sweet chestnut					✓*	✓	
Trazel					✓*	✓	
Walnut					✓*	✓	
Yellowhorn			✓	✓	✓		

* Plants grown from seed may not be true to type. Cultivars of these are usually grafted.

Softwood cuttings

Take these in midsummer using 10-15cm (4-6") of new growth with a firm (not soft and floppy) base. Bottom heat (a heated bench or propagator), rooting hormone and mist are usually required. Rooting usually takes 4-8 weeks and careful hardening-off is needed. Not easy.

Hardwood cuttings

Hazel cuttings should be taken in late winter, using 15cm (6") cuttings from vigorous shoots on stooled (recently coppiced) plants. Needs bottom heat and rooting hormone. Not easy.

Root cuttings

Take these in midwinter, using 3cm (1⅛″) sections of root, and plant horizontally in pots.

Suckers

Dig suckers in winter, and pot up for a year if not well rooted before planting out.

Seed

Nut trees will not usually come true from seed – i.e. the plant you grow will not be identical to the mother tree where the seed came from. Genetically, the male pollinator tree will also be represented in the seedling tree. If both the mother tree and the pollinator are good nut producers, the chances are the seedling will be too.

Species in the hickory family and walnut family can sometimes cross-pollinate. This means that growing seedlings of these trees can lead to hybrids, if the trees came from an orchard where different species are cultivated together. So, for example, butternut and heartnut can hybridize to produce the buartnut, so you may grow a heartnut seed and end up with a buartnut.

Seed of most nut tree species requires 'stratification' or autumn planting.

Stratification means a period of cold, moist conditions – basically persuading the seed that it has been through a winter lying on the ground. Hazelnut, hickory, pecan, trazel and the walnut family (black walnut, buartnut, butternut, heartnut, walnut) all need 3-4 months of stratification. Pine nuts, depending on species, need 1-3 months; yellowhorns 6 weeks; and bladdernuts 6-8 months, half in the warm followed by half in cold.

Seeds that are not dormant require autumn planting. These seeds have a high water content and a thin shell, and don't usually store well in dry conditions. Non-dormant seeds are sweet chestnuts, chinkapins, monkey puzzles and acorns from oaks.

Most nut seed grows with a deep taproot, so should be sown in deep containers or trays. Take care not to break the roots when potting up and use a well-drained compost mix. Protect seed that has been sown from rodents, as they will delight in digging it up to eat! You might also need to protect young seedling trees – I have had rats and squirrels pull up seedlings to get at what is left of the nut.

Seedling plants have the advantage of greater genetic diversity as well as being much easier to grow compared with grafted trees. If you have plenty of space and want to grow on a reasonable scale, one way to plant is to grow a large number of seedlings, plant quite densely, and thin them out as they grow to leave the best-performing plants. I know of a few folk doing this with sweet chestnuts. Seedling rootstocks are also widely used to produce grafted trees.

Grafting

To produce a grafted tree, a piece of wood from the variety you want (the scion) is attached to another piece of wood with roots (the rootstock). They are attached so that the union (where they meet) heals and the plant grows as one. If the rootstock itself is a named cultivar (rather than a seedling), then it is called a clonal rootstock.

Grafting of most nut trees is more difficult than grafting fruit trees because the graft union needs substantial heat to callus and heal properly. The walnut family needs 27°C (80°F) and sweet chestnut about 21°C (70°F). In a cool spring climate these temperatures are difficult to achieve without special equipment to keep the graft unions warm. (Heating up the root system of the rootstock or the scionwood of the cultivars grafted should be avoided, otherwise they will come into growth before the graft has healed.) A 'hot graft pipe' is the easiest to use on a small scale.

Almonds are usually grafted on to almond, plum or hybrid *Prunus* rootstocks, sweet chestnuts are sometimes grafted on to clonal chestnut rootstocks, and there is ongoing work to produce hazel and walnut clonal rootstocks. But otherwise, in cool climates, seedling rootstocks of the same species being grafted are used.

Grafting on to small-diameter rootstocks is generally done in late winter or early spring. Various types of graft are used, including whip-

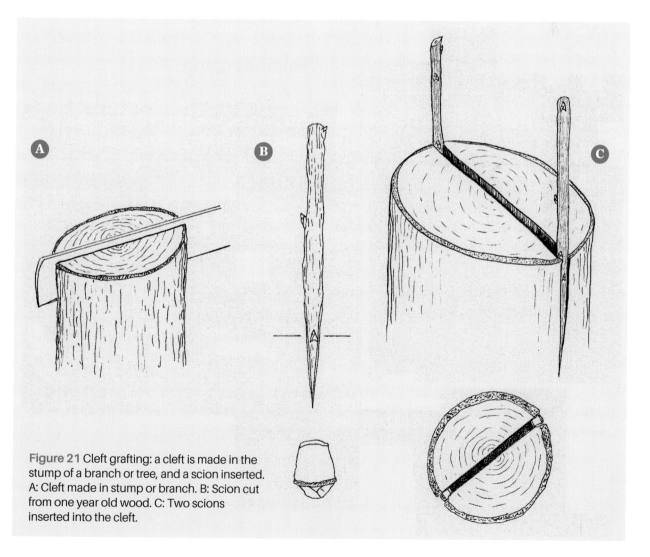

Figure 21 Cleft grafting: a cleft is made in the stump of a branch or tree, and a scion inserted. A: Cleft made in stump or branch. B: Scion cut from one year old wood. C: Two scions inserted into the cleft.

and-tongue and wedge grafts. A whip-and-tongue graft involves cutting a sloping cut with a 'tongue' on both the rootstock and the scion; the tongues then fit together in the graft. In a wedge graft a V-shaped cut is made in the scion and a vertical cut in the rootstock. The vertical cut is eased open with a knife and the wedge of the scion inserted.

The walnut family can be tip-grafted in mid-summer, using a small terminal piece of green shoot and a wedge graft.

Budding is a type of grafting where just one bud is transferred to a rootstock. Most nut trees can be budded in summer – midsummer for walnuts, mid- to late summer for others.

Pines are more difficult to graft, but this can be done in early spring, with side grafts protected inside plastic bags. In Spain, some improved fruiting varieties of stone pine (*Pinus pinea*) are being propagated this way.

Established trees of most nut species can also be regrafted in the field using simple cleft grafting (see Figure 21) or other grafting methods.

Because this is done in mid-spring when air temperatures are warm, even grafts needing heat, like walnut, can work well. The trickiest thing might be getting hold of the scionwood needed to make the graft, as few nurseries sell it. But using this method you can plant a cheap (or free) seedling tree and aim to regraft to a good fruiting variety 3-5 years later.

Should you need more detailed information on grafting techniques, I recommend R. J. Garner's excellent reference book *The Grafter's Handbook* (see Resources).

Stool layering

Stool layering (see Figure 22) involves planting a mother tree on its own roots and then, once well established, annually coppicing it low. The new shoots that grow are girdled with copper wire and mounded up with soil or sawdust through the season. As the shoots grow, the wire induces the shoots to root above the wire. At the end of the season, the rooted shoots are unearthed and cut off, and can be planted straight out or grown on for a year in a nursery.

Hazelnuts are usually propagated on their own roots by the stool layering method. This is because they are naturally suckering trees, and hazels grafted on seedling rootstocks will always be prone to the rootstock shoots taking over. Propagated on their own roots, hazelnuts can then be grown as a multistemmed bush or as a single-stemmed tree.

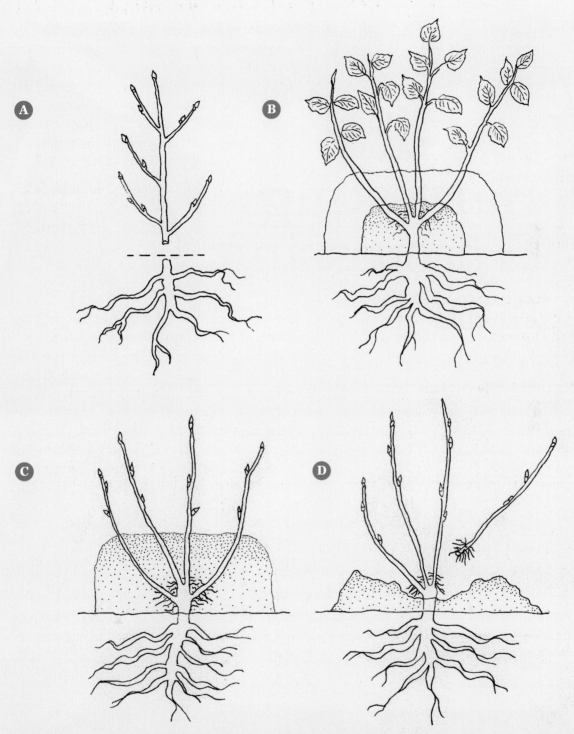

Figure 22 Stool layering of hazelnut. A: Mother tree cut back in winter. B: New shoots mounded up in summer, after girdling with wire. C: Shoots rooting in autumn. D: Rooted shoots removed in winter.

Chapter
THREE

Harvesting & processing nuts

It is worth thinking about how you are going to harvest and process your nut crops at the outset. Are you willing to put in hours of back-breaking work, hand-picking nuts off the ground? Similarly, nuts need at the very least cracking before you eat them – are you really going to hand-crack a big nut crop with an 18th-century nutcracker? This chapter describes harvesting methods for small- and large-scale sites, and recommends well-designed implements that make the harvest quick, easy and fun. It also explains the various stages involved in processing nuts, and gives details of equipment that will save you time and effort.

Harvesting

Nuts are harvested in autumn (in very hot regions, late summer as well).

Although most nuts are harvested off the ground, a few (bladdernuts and yellowhorns) are more likely to be picked off the trees. These are simply harvested when ripe.

For ground harvesting, grasses should be cut low just prior to nut harvest. If the harvest goes on for more than three weeks or so, they may need cutting again.

If grasses are grazed, for example with sheep or geese, then the animals must be removed from the area four weeks before nut harvest begins, to minimize any risks of E. coli infection from the manures. This means that it is likely the grasses will still need cutting with a mower or topper just before harvest.

When choosing perennial plants to grow beneath your nut trees, opt for those that will have died back to underground parts by nut harvest (e.g. to a bulb, like wild garlic) or can be cut without detriment to the plant. An alternative to cutting plants to achieve low ground cover for nut harvest is to harvest using nets – see page 68.

How you harvest will depend on the scale of your nut tree growing. However you do it, try to harvest regularly. In a moist climate, nuts lying on the ground rapidly pick up fungal rots. You can leave walnuts on the ground in California for a week or two, but in cooler climates you should harvest every two days unless the weather is unseasonably dry and warm.

If the weather has been calm with little wind, then if possible shake or knock branches just before you harvest to dislodge loose nuts. Large-scale nut growers use tree shakers, but on a small scale you can shake individual branches of trees or use a long pole to tap them. I use a 10m (33') bamboo pole.

Not all nuts fall freely from their husks or burrs. Some chestnuts fall within the burrs, and have to be extracted – if hand-harvesting, then a kick of your heel opens the burrs efficiently. Some nuts fall within their husks and are harvested with the husks on – for example heartnuts. The husks are removed later.

Harvesting methods

You should consider how you are going to harvest your crop beforehand. Unlike fruit, which is mostly picked off the tree, nuts are harvested from the ground. You can save yourself lots of time and backache by using appropriate tools or equipment.

* **Hand-harvesting.** Picking nuts off the ground, the traditional way, is fine for just a few trees, although quite slow – up to 5kg (11lb) per hour or so. It's also hard on your back – you'll appreciate why French children traditionally got given several weeks off school to help with the chestnut harvest.
* **Nut Wizard or similar hand-held harvesters.** These fantastic harvesters were developed in the USA and are available worldwide. They comprise a long handle and a head of bendy wires. When rolled over the ground, any objects like nuts spring inside the head and are retained there until the head is emptied. There are models designed

The author harvesting walnuts with a Nut Wizard.

Nut Wizards for harvesting acorns, hazelnuts, chestnuts and walnuts.

A push-along nut harvester.

for acorns, hazelnuts, chestnuts and walnuts. I use these for most of my nut harvesting. They don't pick up many leaves or other debris and are three to four times faster than picking nuts by hand.

* **Push-along harvesters.** These look a bit like lawn mowers. They have a rotating cylinder of bendy twines that grab and pick up nuts and deposit them into a hopper. Made in France and the USA, these work very well on nicely flat ground, but if your site has dips and bumps then some nuts will be missed.

* **ATV (quad bike) pulling small harvesters.** The harvesters are similar to the push-along ones described above, with several used together in a series. Useful on a mid- to large-scale planting where the orchard floor is very flat.

* **Tractor-scale commercial harvesters.** There are various types. One is a large-scale vacuum cleaner that sucks up nuts (and leaves, stones, burrs, etc.); others use brushes to sweep up nuts. Brushes require a dead flat orchard floor, and all these harvesters tend to pick up extra material that needs sorting out before the nuts can be processed.

* **Nets.** A few commercial nut orchards use nets, sometimes strung between wires (attached to posts) so the nets are held off the ground. Nets can also be laid on top of grasses/plants. Because the nets have to be manipulated every few days to move the nuts to the edge to harvest, the plants don't get a chance to grow through the netting. Nets will catch burrs and leaves too, of course, so there will be work sorting the mixture out.

Processing nuts

Nut processing after harvest can involve several stages, depending on the nut in question and where the nuts are destined for:

* Husk removal
* Empty nuts, wormy nuts and debris separation
* Sorting by size
* Drying
* Storage
* Shelling/cracking
* Pellicle removal
* Pressing oil from nuts
* Cooking

Husk removal

The well-known commercial nuts (almond, hazelnut, pecan, sweet chestnut and walnut) mainly fall free of husks and can be harvested very cleanly by hand or fairly cleanly mechanically.

Some nuts of chinkapin and sweet chestnut fall still inside the burrs, which are easily opened when hand-harvesting. When machine-harvested these need removing by feeding the nut mixture through de-huskers (usually rollers at fixed distances, enough to squash the burrs open without damaging the nuts inside).

Other nuts – notably black walnut, buartnut, butternut, heartnut and some hickories – fall still enclosed in the husks. Left on the ground, the husks will rot and/or be eaten by slugs, leaving the nuts. But this usually leads to nut kernels being stained black and many will be inedible, so you need to get the husks off quickly. On a small scale they can be rubbed off by hand quite easily. On a larger scale the easiest method of husk removal is to use a concrete mixer! Place the nuts with a few handfuls of stones and some water inside the mixer and set it going – after a few minutes the husks will have been rubbed off. Pour out the mixture and pick out the nuts.

Empty nuts, wormy nuts and debris separation

Empty nuts and bits of leaf and husk can be separated off by floating. Good nuts will sink and empty ones will float. Nuts with weevil larvae feeding inside often float too but not 100 per cent. Hazelnuts have the highest proportion of empty nuts – up to 30 per cent – due to their pollination quirks (see Part II, page 151) and are always worth separating via floating.

Sweet chestnuts grown commercially are often given a hot-water soak (50-68°C/122-154°F for 45-50 minutes). This kills any chestnut codling moth larvae inside the nuts (see Part II, page 233), to stop them moving from nut to nut in storage, because these nuts are stored fresh. For nuts that are dried (i.e. most), any larvae inside nuts will be killed by the drying process.

Sorting by size

On a home scale you don't need to sort nuts by size. But on a larger scale, whether the nuts are for you or to sell, sorting becomes more important. All mechanical nutcrackers work best when the nuts are all about the same size, so they are adjusted to work for that size. And if you are selling nuts, the public don't like buying small nuts so these should be removed (and preferably used or processed yourself).

Our chestnut sizing box. Small nuts fall through the holes and are removed by opening the removable front panel.

Walnuts (a double-size variety, Cobles #2) in an Excalibur dehydrator, one of the best commercial small-scale dryers.

Most nut sorting is done by making nuts roll over a horizontal surface with holes of a specific size. Commercially there will be a large vibrating surface with small holes highest and larger holes further down, so that several different sizes (grades) of nut can be sorted all in one go.

On a smaller scale it is easy enough to make a box sorter. I use a home-made sorter (see photo above) to sort sweet chestnuts, using a hardboard surface drilled with 26mm (1") holes. The nuts are moved around by hand over the surface and small ones drop through. The larger ones can be sold and I use the smaller ones to dry for my own use.

Drying

If you grow nuts in a moist climate, then you will need to dry them to store them, otherwise they will become mouldy within a week or two. Even in drier climates, most nuts dried in ambient air will store for 1-2 months at most. It is best to dry nuts in shell because cracking only works well with dry nuts, and because dry nuts in shell store much better than shelled kernels.

The two things needed to efficiently dry (dehydrate) nuts are warm air and air circulation. Warmth alone is not sufficient, as the still air soon becomes saturated with moisture from the wet nuts. Air circulation alone is also not usually sufficient. All that happens in these cases is that either the cool air or the warm, moist atmosphere encourages fungal rots, which can very quickly ruin crops.

If you live in a continental climate, where warm sunny autumn days are the norm, a solar dryer is certainly possible. There are lots of designs on the Web. But without full sun, you are wasting your time solar drying (I've tried it!). In cool temperate climates, a good-quality bought dehydrator, such as the Excalibur, or a home-made dryer will give the best results.

The drying process

Arrange the nuts on drying trays, leaving small spaces between the nuts for air circulation. Different nuts can be dried together, as they don't have a strong flavour or odour. Try to interrupt the drying process as little as possible. Don't add fresh nuts to a dryer filled with partially dried nuts – the increased humidity will greatly increase the drying time of the partly dried nuts.

Test frequently near the end of the drying process to avoid overdrying. You can use the moisture content and drying information from the table below, and weigh the fresh and dry nuts to get to an accurate dryness.

For example, for sweet chestnuts:

❋ Weigh the fresh nuts, e.g. 5kg (11lb).
❋ The final dried weight (to achieve 10-15 per cent) is about 47 per cent of 5kg, i.e. 2.35kg (5lb 3oz).
❋ So keep drying until the weight reaches 2.35kg.

Nuts should be dried as soon as possible (within 24-48 hours) after harvesting. Drying temperature is important, especially for oily nuts (e.g. almonds, hazelnuts, pecans and walnuts) – don't dry at warmer temperatures than those listed below, otherwise you run the risk of turning the oils rancid.

Starchy nuts like chestnut are high in water content and take a substantial time to dry, so they can be stored long term. Oily nuts have a much lower water content and take less time. The exact water content will depend largely on weather conditions around harvest. If it has been raining on to nuts that are sitting on the ground, they will have absorbed water and their initial water content will be nearer the top end of the variation listed below.

Nut drying data

Nut	Initial water content	Drying requirements	Dry weight as a % of fresh weight (in shell nuts)	Dryness test/comments
Almonds	15-20%	Dry to 10% or less at 25-40°C (77-104°F): may take 8-15 hours Dry shelled almonds to 6%.	82-87 (drying to 10%)	For mechanical shelling, 8-10% is needed.
Chestnuts	60%	Dry to 10-15% moisture at 40-50°C (104-122°F): may take 3-5 days. Dry to 7% moisture for making flour.	46-49 (drying to 10-15%) 44 (drying to 7%)	15% is ideal for chestnuts to be eventually rehydrated and gives storage for over a year. Drying to 7% moisture or less is needed to make good chestnut flour which doesn't cake – may take a further 2-7 days.
Hazelnuts	25-35%	Dry at 40-50°C (104-122°F) to achieve 7-8% moisture. Typically takes 20-40 hours.	73-81 (drying to 7-8%)	
Pecans		Dry to 8% moisture: may take 20-40 hours.		Shelled kernels are dried to 4% moisture.
Walnuts	25-35%	Dry to 12% moisture at 25-43°C (77-110°F). Typically takes 20-40 hours. Dry shelled kernels to 8%.	70-78 (drying to 12%)	Membrane between the two walnut shell halves is crisp and not rubbery.

Home-made drying cabinets

You can make your own drying cabinet fairly easily. I have two, constructed out of plywood 12mm (½") thick (see Figure 23).

You need a hole at the bottom for a fan to draw air in, and an outlet hole at the top. I use an axial plate fan rated at 37m³/min (1,300ft³/min) for a cabinet 1.8m (6') high. You need a system of trays to layer nuts with air spaces between – we use standard 60 x 40cm (24" x 16") food-grade plastic stacking trays which have air gaps in the base. These can stack on top of each other, meaning you don't need rails to slide trays in and out. For my cabinet, these trays sit on a standard dolly with wheels, so I can wheel the whole stack of trays in and out to check them. As a heat source I use an ordinary fan heater with two settings (1kW/2kW) that sits at the bottom of the cabinet beneath the trays in the airflow created by the main fan. The temperature this creates depends partly on ambient air temperature but is between 30 and 40°C (86-104°F), which is ideal for drying nuts.

The nuts are placed in shallow layers (only two nuts deep in our trays, as the trays themselves are quite shallow) so that the airflow can travel through them and eventually out of the cabinet. The moist air is vented outside using flexible plastic ducting (there is potential for venting this humid, moist air into a greenhouse to use some of the warmth).

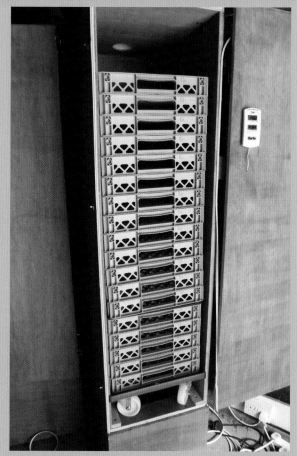

Stacked trays in our drying cabinet.

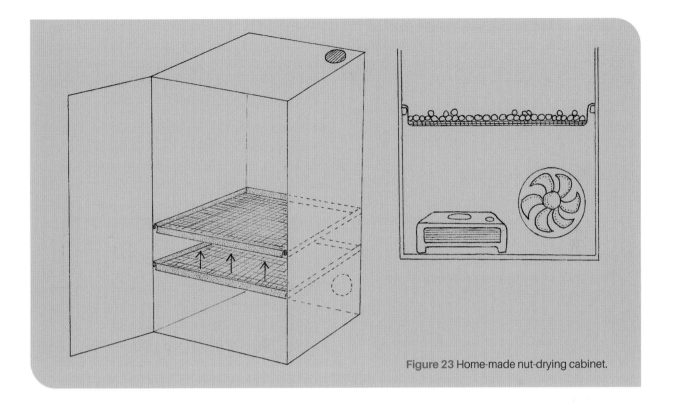

Figure 23 Home-made nut-drying cabinet.

Storage

Only store nuts that are in good condition. In most nut crops, the edible kernel grows mainly in the last few weeks before harvest, and if this growth is exceptionally fast, then in nuts with thin shells (e.g. sweet chestnuts), the kernel can outgrow the shell and cause the shell to split. The kernels are still perfectly edible at harvest time, but they must be eaten or utilized quickly, as they will mould within a week or so.

Some nuts with a high water content can be stored fresh at low temperatures (0-5°C/32-41°F) with some airflow and high humidity for a few months. On a home scale, place nuts (surface dry) in nets and store in a fridge. Be vigilant in watching out for mouldy nuts. On a commercial scale, refrigerated stores are used.

Of dried nuts, starchy nuts (e.g. sweet chestnuts) store for longer than oily nuts (e.g. walnuts) because the oils turn rancid in time:

* Dried shelled starchy nuts will store for five years or more.
* Dried oily nuts will store for three-plus years (in-shell) or one year (shelled).

The storage life of dried nuts is about 5 times longer when stored at 0°C (32°F) than at 15°C (59°F), and 10 times longer when stored in a freezer at -20°C (-4°F).

Good storage of dried nuts depends on airtight, moisture-proof packaging and dark conditions. On a home scale, storage in large self-seal plastic bags suffices. Commercial growers usually quickly sell their crop and are not concerned with long-term storage.

Vacuum packaging also extends the storage life of dried nuts – there are small-scale machines for home use available from dehydrator suppliers (see Resources).

Obviously, take precautions against rodents when storing nuts!

Shelling/cracking

The time consumed in cracking nuts before eating can be off-putting and a decent cracker can make a huge difference to the quantity of nuts you end up eating! See Resources for companies that make crackers and cracking machinery.

Some nuts have soft shells when fresh (e.g. sweet chestnuts) that are sometimes peeled off prior to cooking. Chestnut-peeling pliers (or any spring-opening pliers) peel a nut at a time with reasonable speed. Alternatively, chestnuts can be peeled after cooking, which can be easier.

Small chestnut-peeling machines have recently been developed by New Zealand nut growers and will be available commercially. Dried chestnuts have brittle shells and can be shelled like other dried nuts, as below.

To begin with, the nutshells must be dry to crack adequately. If you have no cracker, then you can at least half fill a sack with dried nuts and beat them before sorting the mixture out. You might find some varieties shell more cleanly than others – mostly due to the inner surface of the shell and how rough or convoluted it is.

There are a variety of hand-held nutcrackers of different designs. These are designed for the main nuts – hazelnuts and walnuts – and will handle heartnuts. But they will not handle hickory nuts or other walnut family members (black walnut, buartnut, butternut) – in fact using these nuts usually breaks the cracker!

Hand nutcrackers vary widely. The conical squeezing type is generally good. Blade-type openers are made for walnuts. The dog cracker here is interesting but not particularly good!

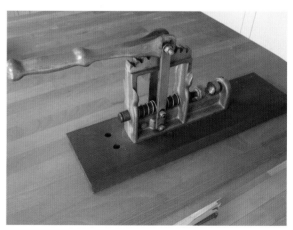

This American cracker copes with the toughest nuts.

For the tough nuts that standard hand crackers won't cope with, you need a much heavier-duty mechanism like the two below.

A few hand-cranked nut-cracking machines are available commercially. These greatly improve cracking speed and are useful even for a small grower.

There are also a few electric small-scale nutcrackers, such as the one on the right.

There seem to be few mid-scale cracking machines around, although large-scale equipment is available (see Resources). For mid-scale machines – for example, if you are going to grow a few acres of nuts – you might end up doing like I have and making your own machines (see box, page 76).

This Turkish cracker works very well and is currently my default kitchen cracker.

Hand-cranked nut-cracking machines from Europe (Mycronut) and the USA (World's Best).

Home-made nut-cracking and sorting machines

The dearth of commercially available mid-scale nut-cracking technology induced me to build my own machines.

The nutcracker

My starting point was a promising design by Harry Lagerstedt, published in a 1992 issue of *The Nutshell* magazine in the USA.

The two important parts of the machine are a rotating metal wheel and an 'anvil' (see Figure 24). The heavier the wheel the better, since the more mass it has, the less likely it is to slow down as nuts are fed in. The wheel in our machine is 5cm (2") wide and 25cm (10") in diameter; too small a diameter will not work well. The wheel has 8 beads, about 2mm ($^5/_{64}$") high, welded across its surface. These are placed not straight across but are V-shaped, so that nuts are funnelled towards the centre of the wheel as they are cracked, and are less likely to travel to the edges of the wheel, where pieces might conceivably get trapped between the outer side of the wheel and the casing.

The wheel shaft is mounted in two bearings and fitted with a large pulley, and a belt connects this to the electric motor. The motor itself is second-hand, having been salvaged from an industrial paint-mixing machine, and runs at about 2,000 rpm. Hence the pulley ratios were made so that the cracking wheel rotates at about 400 rpm, which seemed, by trial and error, to be the most suitable cracking speed.

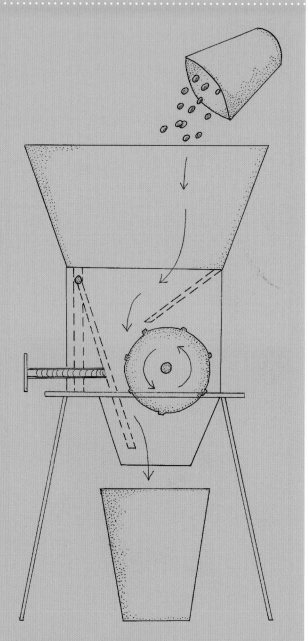

Figure 24 Nut-cracking machine.

The anvil is the hard place against which the nut is squeezed. It is hinged so that the space between the anvil and wheel can be adjusted via a threaded bolt that can be moved against the anvil, pushing it towards the rotating wheel. This is the adjustment used for cracking different-size nuts. The anvil is made from 10mm-($^3/_8$")-thick steel, and is hinged at its top with a 10mm-($^3/_8$")-diameter steel pin going through two heavy metal plates welded to the frame. This pin can be easily removed if any nuts get jammed, and the anvil opened to its widest to allow access into the cracking zone.

The author's nut-cracking machine. You can see the L-shaped anvil pin and top of the anvil.

The whole mechanism is mounted on an angle-iron frame and is enclosed by a sheet-metal housing. The lower part of the wheel and anvil are also enclosed to guide the cracked nuts into a container. A sheet-metal hopper on top of the housing holds the nuts prior to cracking. Throughput of nuts is impressive. I've measured walnuts being cracked at about 45kg (100lb) per hour and hazelnuts at 90kg (200lb) per hour. The quality of the resulting cracked nuts is dependent on a fairly uniform size, so much better results are obtained when nuts are graded first. Graded hazels can be cracked with about 95 per cent of whole kernels; ungraded hazels come through at 85-90 per cent. Walnuts come through with a good proportion of halves. Dried chestnuts go through the machine too, with kernels broken into small pieces (small enough to go through an electric grain mill) with whole or large parts of skins; the cracking rate is intermediate between hazels and walnuts. Once you have a fast nutcracker, the time-consuming process becomes separating the shells from the kernels. Inclined boards, pneumatic (air-blown) separators and vibrating grading tables are all possible answers. We built our own nut sorter.

The nut sorter

If you have a nut-cracking machine that works at a good speed, then the next problem is to efficiently separate the kernels from the bits of shell. Commercially, most sorting at this stage uses a column of moving air in a machine called an air leg aspirator. When the kernel/shell mixture is introduced into an airstream of the correct velocity, most bits of shell are blown clear, while the heavier kernels remain in the airstream. The basic design is quite simple (see Figure 25). A fan blows air through a chamber of specific dimensions. The fan must be centrifugal, not axial – i.e. it doesn't have a propeller, but instead has straight blades that throw air away from the blade tips. This type of fan produces an even airflow across the whole width of the fan outlet, whereas an axial fan doesn't and would lead to problems of vortices, etc. in the separating chamber.

My centrifugal fan is in fact sold as a carpet dryer and can be found quite easily for sale under that description. It has three speed settings and requires between 680 and 900 watts of power to operate.

A mesh near the bottom of the chamber is used to stop nut parts falling into the fan. The chamber itself has a width of 240mm (9½") and depth of 125mm (5"). It has a curved top part to eject shells. The distance from mesh to the lowest part of the exit mouth is 880mm (34½"). The whole machine is made out of steel, but for commercial use, the parts in contact with the kernels would need to be made out of stainless steel.

Cracked nuts – kernels and shell parts – are loaded into the sorter through a hinged flap midway up the back of the machine. These immediately fall down on to the mesh to await sorting. If too much material is loaded, then air won't flow past it properly and lift it – the limit is about a 1kg (2lb 2oz) of material at a time.

The machine is switched on and, if necessary, set to the correct speed. Shells are steadily ejected

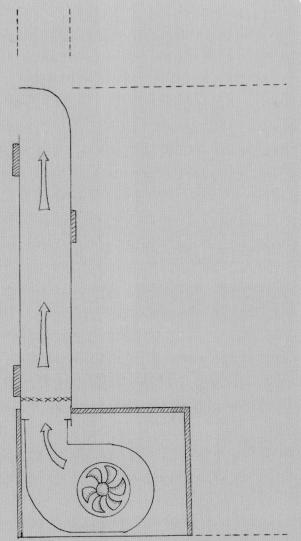

Figure 25 Nut-sorting machine.

from the top – our machine is usually placed in a shed doorway to eject straight outside. Sorting of a batch is completed in about 60 seconds. The machine is switched off and clear kernels are then removed via the lower hinged flap, which is level with the mesh inside the separating chamber. An adjustable vent was added on the front of the machine in case of nuts being sorted that needed a lower chamber height.

Efficiency of sorting is about 90-95 per cent, with very few tiny kernel pieces ejected, and a few pieces of shell remaining. Hazelnuts, which tend to crack more uniformly, are at the higher end of the scale. If needed, walnuts can be double-cracked – once on a normal cracking setting for walnuts, then a second time on a hazelnut setting. This second cracking is to crack half walnut shells into smaller pieces – it doesn't really break the kernels any smaller – because complete half walnut shells don't separate well.

Hazelnuts get well sorted at the medium and high settings on our machine (fan air delivery of 130m³/min [4,600ft³/min] and 150m³/min [5,300ft³/min] respectively). Walnuts only sort well on the higher setting. A fan with higher air delivery will require a larger cross sectional area of separation chamber and vice versa.

Nut-sorting machine. Note walnut shell just being ejected at top, and the closed lower flap.

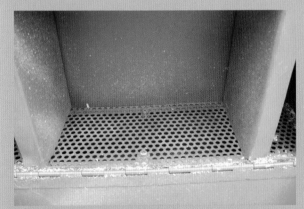

Lower flap opens to show collecting chamber with mesh at the bottom.

The centrifugal fan.

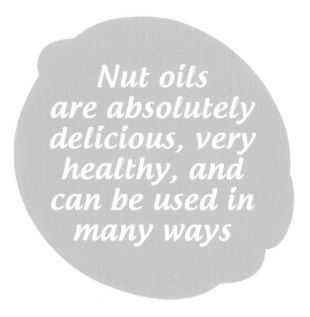

Nut oils are absolutely delicious, very healthy, and can be used in many ways

Pellicle removal

All nuts have a thin papery inner shell (the pellicle) surrounding the kernel, which is usually bitter when fresh, although once dried the bitterness largely or wholly disappears. The amount of bitterness can vary a lot between cultivars.

When nuts are fresh, the pellicles are not particularly easy to remove, but once dried, they rub off easily. Commercially, nuts like almonds are dried, cracked, and then the kernels are tumbled together to rub off the papery pellicles, which are then blown off.

Sweet chestnut pellicles determine whether the cultivar is a *marron* or a *châtaigne* (see Part II, page 238), and the best cultivars for fresh eating are *marron* types, whose pellicle is easily peeled off once the nuts are cooked.

Pressing oil from nuts

Nut oils are absolutely delicious, very healthy, and can be used as a salad oil or in cooking in many ways. As a speciality oil they are expensive to buy, but if you grow your own oily nuts, then making your own oil is a natural extension of using the crop.

After pressing the oil from some nuts, the left-over 'cake' can also be used, particularly in cooked dishes.

Unfortunately, there is not much small-scale oil extraction equipment around. The Piteba oil expeller was developed for use in poorer parts of the world – it is hand-cranked, cheap, and expects the user to fashion a hopper out of half a plastic bottle to hold the kernels. Our experience is that unless you mount it securely, it is difficult to use and easily broken. It is quite fiddly, and requires a small burning flame to warm the kernels before they are crushed to expel the oil.

There are various inexpensive small-scale electric oil presses/expellers available from China, which are worth considering if you want make your own oil (see Resources).

For commercial-scale growers, stainless steel oil presses are available (see Resources).

Cooking

Methods of cooking and use depend on whether the nuts are predominantly starchy or oily.

Starchy nuts

Chinkapin, ginkgo, golden chinkapin, monkey puzzle, oak, sweet chestnut, yellowhorn.

Like other starchy foods, starchy nuts are usually always eaten cooked. Nutritionally they are very similar to normal grains (wheat, rice, etc.) and contain very little fat. They can be boiled, roasted, etc. These starchy nuts can also be dried and then ground into flour (gluten-free), which can then be used to make flatbreads, pancakes, biscuits, cakes, or mixed with cereal flours in breads, and so on.

Oily nuts

Almond, black walnut, buartnut, butternut, hazelnut, heartnut, hickory, pecan, pine nut, trazel, walnut.

Oily nuts are rich in healthy oils, and there is increasing evidence that the oils particularly aid the healthy functioning of the cardiovascular system as well as having other benefits. In some countries, for example Australia, oily nuts are included in official recommended daily dietary guidelines.

These nuts are all good to eat both raw (as they are, in salads, muesli, etc.) and cooked (in many cooked savoury and sweet dishes). They can also have the oil pressed from them to use as a high-quality culinary oil. These oils have a relatively short shelf life and should be kept in cool, dark conditions.

Readers will be aware but it is worth remembering that the protein in many nuts can cause an anaphylactic reaction in sensitive individuals.

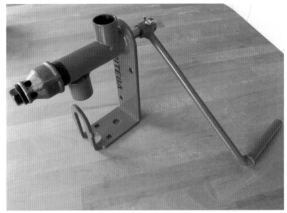

The Piteba oil expeller.

A Chinese electric oil expeller. Nuts are fed into the screw mechanism, warmed with an internal electric element and the oil squeezed out. The 'cake' left over can be used in cooking, etc.

Part TWO

Nut trees A–Z

ALMOND
(Prunus dulcis)

ZONE 6-7, H5-6

The almond has been cultivated since ancient times. Originally from Central Asia, it was disseminated to civilizations across Asia (2000 BC), Europe (350 BC) and north Africa (AD 700). Almonds were introduced into North America (California) during the Spanish missions period (1769-1833), but significant plantings were made there only after the gold rush (1848).

Almond fruit

Almond is a very important worldwide commercial crop. Global production of almonds is concentrated in three regions: Asia (Iran, Turkey, Syria, Iraq, Pakistan, Afghanistan, north-west India – all grown by traditional practices; also Israel, Tajikistan and Uzbekistan); the Mediterranean (Italy, Spain, France, Greece, Portugal); and California. Total world production is around 300,000 tonnes per year, with California accounting for 70 per cent of the world almond production (though the 2015 drought caused severe problems for some growers).

The almond is a small deciduous tree growing 3-10m (10-33') high, depending on rootstock and climate (occasionally more), upright branching when young, and of bushy habit with a broad crown when older. Compared with some other nut trees, the expected lifespan is relatively short, often 30-50 years, although some trees can live to 100 years of age or more.

Leaves are long-pointed and finely toothed, 7-12cm (3-5") long and 2-4cm (¾-1⅝") wide, light green above and shiny.

White or pink flowers 2-5cm (¾-2") across are produced singly or in pairs usually before the foliage. Because of the low winter chilling requirement of the tree, almond flowers very early (midwinter to early spring, depending on the selection and locality) as soon as it has experienced enough winter cold.

The oblong fruit is 3-6cm (1⅛-2½") long. It consists of a leathery hull within which is a single nut with a pitted shell that contains the kernel (seed); the hull has a downy exterior, and the tough flesh splits at maturity to expose the pitted shell. The nut can be thick- or thin-shelled, is flattened and brown, and has variable size range: 22-45 x 15-25 x 10-20mm (1-1¾ x ⅝-1 x ⅜-¾"). Within the shell is a single flattened kernel, 20-30 x 12-16 x 5-12mm (¾-1⅛ x ½-⅝ x ⅛-½"). Fruit occur mainly on short spurs.

Sweet almonds – the type normally grown to eat – are from the botanical variety *dulcis*. Bitter almonds, from var. *amara*, have nuts that are bitter to the taste and poisonous to eat – they are mostly grown for oil production.

Almonds are fairly cold-hardy – about as hardy as peaches: normally regarded as hardy to Zone 7 / H5 but some selections are hardy to Zone 6 / H6; the limiting factor in growing it is the frost-sensitivity of the flowers and young fruit.

Uses

Almonds are a concentrated source of energy, supplying significant amounts of fats, protein and fibre (see Appendix 1).

Almonds are delicious raw or cooked, used in sweets, baked products and confectioneries. Nuts are available in-shell, shelled, blanched, roasted, dry-roasted, as a paste (marzipan), a butter, and cut or chopped into various shapes.

Almond oil from the kernels is sweet and scented, valued for marinades and cooking. It is also used as a flavouring agent in baked goods, perfumery and medicines; and used for cosmetic creams and lotions. Sweet almond kernels contain 44-55 per cent oil; of this 67 per cent is unsaturated oleic.

Green (unripe) almonds are used in early summer, when whole fruit are picked before the shell forms and in which the kernel is very soft and tender. These are popular as dessert in almond-growing countries. They are sometimes preserved in sugar or the young fresh kernels in alcohol.

In France, milk of almonds (*sirop d'orgeat*) is a refreshing drink made from crushed almonds that is regarded as having medicinal properties. Almond milk is sold in health-food shops and supermarkets.

Secondary uses of almond

The gum that exudes from damaged stems is edible. It has also been used as a glue.

The seeds and the oil have long been used in traditional medicine for their therapeutic properties, for example diuretic, emollient, laxative, sedative, stimulant and tonic. Used in cancer remedies and for asthma, skin complaints and ulcers, to name a few. The leaves are used to treat diabetes in Chinese medicine. The oil is often used as a carrier oil in aromatherapy.

The twigs and branches have been found to have a repellent effect on flies (including the house fly) and on human head lice.

A majority of almond oil is produced from bitter almonds, as is almond essence for flavouring only. Bitter almond kernels contain 38-45 per cent oil.

The dry fibrous hulls left over after nuts are hulled are a valuable livestock food in commercial almond-growing regions – they contain 25 per cent sugars.

The oil from seeds is an excellent lubricant, used in delicate mechanisms such as watches.

Several dyes can be obtained, including green from the leaves, dark grey-green from the fruit, and yellow from the roots and leaves.

The burned shells were formerly used as an absorbent for coal gas.

The timber, usually only available in small amounts, is used for fuel, carving, cabinet making and turnery. It is purplish-brown, strong and durable.

Almond is a good early bee plant, particularly for bumblebees, being a source of nectar and pollen.

Cultivation

Almonds are adapted to drought and poor soils, and were traditionally grown on marginal soils without irrigation, because they could survive and produce under such conditions. Californian growers initially copied these tactics, but soon found that almonds grew well on fertile, deep, well-drained soil, and responded to irrigation and fertilization so that yields increased by 100-200 per cent. More intensive cultivation nearly always makes trees more susceptible to pests and diseases, in addition to using water resources and fertilizers (usually oil-based). If you want a low-input crop that still crops fairly well in dry regions, almonds are ideal grown in the traditional way.

If almonds are grown near peaches, the flowers can hybridize and produce bitter nuts – so keep peach trees well away.

A Californian almond orchard.

In general a sheltered, sunny site is required both to protect the early flowers (and ensure pollination) and to ripen nuts. All frost-susceptible areas must be avoided.

In cool regions, a south-facing sunny site is essential. However, planting on a north slope can be useful in warmer summer regions to delay flowering in spring.

Tree spacing depends on the rootstock used and fertility of the soil. For vigorous rootstocks (almond and peach seedling) a spacing of 7.3m (24') is normally used, while for moderate rootstocks ('St Julien', etc.) 6m (20') spacing is used.

Orchard plantings are usually square or hexagonal (triangle or diamond). A hexagonal planting has all trees equally distant from each other and allows for about 15 per cent more trees per unit area, and increased production per unit area in the early years (see table below).

Distance between trees	Number of trees - square planting	Number of trees - hexagonal planting
6m (20')	277/ha (111/acre)	318/ha (127/acre)
7.3m (24')	187/ha (75/acre)	215/ha (86/acre)

Many commercial orchards are managed by mowing a ground cover of annual weeds. Nitrogen-fixing cover crops like clover (white or subterranean) can also be cultivated and mown. Around the trees a strip 2m (6'6") wide is kept weed-free (using herbicides by most large-scale growers, though mulches can be used by organic growers).

Rootstocks and soils

Almonds are normally propagated on rootstocks that are suited to a range of different soil conditions. The main rootstocks used are:

Almond seedling

Produces vigorous, deeply taprooted trees that require a well-drained soil. Well adapted to drought and calcareous conditions; poorly adapted to wet and waterlogged conditions. Susceptible to crown gall (*Agrobacterium tumefaciens*), honey fungus (*Armillaria* spp.) and crown rot (*Phytophthora* spp.). Also very susceptible to nematodes. Slow-growing when young (delaying bearing), but with age trees can become large and very long-lived. 'Mission' seedlings are sometimes used in California.

Peach seedling

Produces vigorous, somewhat shallow-rooted trees requiring moderately or well-drained soil. Well adapted to drought and slightly acid soils; poorly adapted to drought and calcareous soils. Trees become productive more quickly than on almond seedling. Susceptible to the same diseases and pests as almond seedling. Widely used in California where 'Nemaguard', 'Nemared' and 'Lovell' seedlings are used.

Peach-almond hybrids

These are well adapted to drought and calcareous soils, produce quickly and heavily, and are long-lived. Trees have deep, well-anchored vigorous root systems. Selections include 'Adafuel', 'GF557', 'GF677', 'Hansen 536', 'Hansen 2168'.

Plum rootstocks

Many plum stocks have quite good compatibility with almonds, and are mainly used in wetter summer regions, as they tolerate wetter soils, but lack drought tolerance. Plum rootstocks make almonds hardier. Most of these rootstocks dwarf trees to about 65-70 per cent of the size on peach or almond stocks. 'Damas', 'GF8-1', 'Ishtara' and 'Marianna 2623' are sometimes used along with the following:

* 'Krymsk 86': A peach-plum hybrid stock of low to medium vigour. Tolerates poor soils and grass competition.
* 'Marianna 2624': A moderately vigorous stock. Tolerates heavy wet soils; resistant to honey fungus, nematodes and crown gall. Some cultivars incompatible.
* 'Myran': A vigorous plum–peach hybrid developed in France. Tolerant of wet soils and honey fungus, resistant to root knot nematodes. Good compatibility.
* 'St Julien A': A moderately vigorous stock. Tolerates some drought and waterlogging, resistant to honey fungus. Good compatibility.

Pollination

Although early-flowering almonds can flower from midwinter, late- and very late-flowering

Almond flower.

cultivars flower in early spring. From flowering time onwards, temperatures below -3°C (27°F) can seriously affect fruiting.

Cross-pollination is essential for most cultivars. At least two, preferably three, compatible cultivars should be grown together in orchard plantings, using whole rows of each variety alternately. For a particular variety, the earlier bloom usually has a higher percentage set than does the later bloom on the tree. Hence a main crop variety is better pollinated with another that is flowering slightly before it than with one flowering just after it. Several self-fertile cultivars have been bred, with more recent ones being better suited for large-scale cultivation than older ones.

Temperatures of 15°C (59°F) or more are needed for good pollen shedding. Pollination is heavily dependent on bees, and good crops require plentiful nearby bee activity. In warm regions, hives (3-8 per hectare / 1-3 per acre) are placed in almond orchards at flowering time. Warm, sunny weather immediately after flowering increases both nut set and nut size.

In cooler regions, flowering occurs very early in spring, before hive bees are likely to be flying much. Here, wild (bumble) bees are more likely to the pollinators, and these should be encouraged wherever possible. Artificial pollination, by hand using a camel-hair or rabbit-tail brush, may be needed in cool regions on a small scale; when attempting this, all flowers should be pollinated if possible.

Flower induction for the following year takes place in late summer. On bearing trees, flower buds typically develop on short spurs 5-10cm (2-4") long, which extend each year from the

Almond flowers on short spurs.

vegetative terminal bud. Flower buds are formed in the axils of leaves.

The typical 'June drop' usually occurs in mid-spring, when unpollinated and unfertilized fruitlets drop off.

Feeding and irrigation

Almonds are well adapted to poor and impoverished soils, and nut yields with no extra feeding can reach 50-60 per cent of those of heavily fertilized trees. The latter also become much more susceptible to pests and diseases. Hence on a home scale, occasional additions of compost and mulches should suffice for feeding.

Increasing nitrogen inputs by a moderate amount increases crop yields of almonds on many soils. This may be achieved by any of the normal organic fertilizers, or by using nitrogen-fixing plants, etc.

Almonds don't need irrigation to grow and produce crops, but irrigation in dry regions can double yields. In European conditions, irrigation is rarely necessary – even in Spain only 5 per cent of orchards are irrigated in summer and early autumn. But irrigation is widely used in Australia and sometimes in North America. This is not to say that newly planted trees should be left to their own devices in a very dry spring or summer – as for all young fruit trees, young almonds should be irrigated if soils get very dry.

Pruning

Formative pruning and the pruning of bearing trees is the norm in commercial orchards. In damp climates, pruning is best minimized to avoid fungal diseases and bacterial canker, and pruning in winter avoided completely – only prune between late spring and early autumn. Any large cuts should be treated with a protective paint or biological control (*Trichoderma viride*).

The pruning described below applies to open-grown trees. In cool regions, wall-training almonds is possible, using a fan-training system. Fan-trained almonds are vigorous, so regular summer pruning will be required to keep growth under control. Use the same techniques as for wall-training other members of the *Prunus* family (plums, peaches, etc.).

Formative pruning

In the spring after new trees are planted, their tops should be pruned back to 90cm (3') from the ground (unless already feathered above this height).

During the first year, select the main limbs that will form the framework of the tree. These should be selected with as much space between them as possible – ideally 3 branches radiating at 120° (viewed from above) around the main trunk. If possible, one of these should face into the wind. The angle at which limbs are attached to the trunk should be near to 45° – if too vertical, the limb is likely to split.

Between mid-spring and late summer, pinch out the tips of all other branches arising from the trunk, except for the main limbs; in late summer, cut out all these other branches back to the trunk. Just the main limbs and any small branches arising from them are left growing. This method of pruning (called 'long pruning') allows the tree to develop a natural branching habit; fruitwood develops quickly and the canopy, relatively uncontrolled, grows rangy.

Long-pruned trees sometimes need their main limbs tying every few years to reduce the risk of breakage: a rope is placed as high as is possible while encompassing all the main limbs.

During the second and third seasons, apply the same strategy as in year one, moving a level higher in the tree. Aim to create a full goblet-shaped canopy.

Pruning bearing trees

The pruning required in cool temperate climates is normally much less than in warmer climes. Once the head is shaped, occasional thinning, and removal of dead and crossing wood, should suffice.

Almonds produce most fruit on spurs that grow a little each year, typically living for about five-years. In warm climes, very overcrowded trees producing little new wood should have 10-20 per cent of the fruiting wood removed every year: this usually involves 4-6 thinning cuts of 4-8cm-/1⅝-3¼"-diameter branches of older unproductive wood, which is removed back to a lateral which is younger and more productive. These branch cuts are usually painted to prevent infection. A few well-spaced water sprouts, which will bend out when they start to crop, are left to develop new fruitwood. Some thinning of wood in the centre of the tree may also be needed to allow light penetration. Any dead wood should also be cut out. Annual pruning is best carried out immediately after harvest.

Peach twig borer moth.

Pests

- ❋ **Squirrels** may seriously damage growing and mature nuts. See Chapter 2, page 53 for control measures.
- ❋ **Birds: crows, magpies, starlings, blackbirds and woodpeckers** may all eat almonds and damage trees. Bullfinches can sometimes be a problem, damaging fruit buds – frightening by using sound or visual cues is most appropriate. There are numerous bird pests in Australia. See Chapter 2, page 56 for bird-scaring measures.
- ❋ **Navel orangeworm (*Amyelois transitella*).** This pest in North America is a principal cause of wormy kernels. The adult moths lay their eggs on fruit or twigs, and the larvae infest the developing nuts, which can be totally consumed. Some cultivars are less susceptible. The pest overwinters in mummified nuts on trees, so the best control is removal of these after harvest. Two parasitic wasps are also used in California to provide some control.
- ❋ **Peach twig borer (*Anarsia lineatella*).** This is another serious almond pest found in North America, the Mediterranean region and some other parts of Europe. Adult moths lay eggs on fruit and leaves; the larvae feed on leaves, buds and shoots. Hibernating larvae overwinter under the thin bark in young limb crotches on wood that is one to three years old. Mainstream growers often use powerful insecticides against this pest, but sounder alternatives include encouraging bat populations (which will avidly eat the moths) and the use of the biological control BT (*Bacillus thuringiensis*) at pink bud and full flower.

Peach leaf curl damage.

Diseases

- **Leaf curl (*Taphrina deformans*).** Better known as peach leaf curl, this fungus affects leaves that curl and distort, falling off to be replaced with new leaves often unaffected. Worst in humid climates and damp spring weather. Copper fungicides applied as leaves swell and just before leaves fall control the disease; a better strategy is to choose resistant cultivars. Wall-grown trees with protection from rain for two months in spring avoid infection.

- **Brown rot blossom blight (*Monilinia laxa*).** A fungus that infects flowers in damp weather. The flowers wither, and the disease can spread to the shoot or spur where a canker is formed – the shoot often dies back to this point. Wind and rain splash spread the spores. Occurs in most regions; worst when rain or fog are frequent during flowering.

- **Shot hole (*Stigmina carpophila* or *Wilsonomyces carpophilus*).** A fungus that causes lesions on leaves and fruit, leading to small holes in leaves (hence the name). Occurs in most regions; worst when there is frequent and prolonged spring rainfall. An early infestation may lead to heavy shedding of leaves and fruit drop. Not usually serious enough for action to be taken, though a copper fungicide in early spring can be used for control.

- **Bacterial canker and blast (*Pseudomonas syringae*).** A bacterium that affects most *Prunus* species – almond is less affected than other fruit trees. The disease causes isolated cankers, which cause branch death; more serious on trees grown in light sandy soils, on weak-growing trees, on young trees and in wet regions. Blast causes flowers to blacken and young shoots to die back. Blast (found independently of canker) is associated with cold or freezing temperatures at flowering time. Diseased or dead shoots should be cut out in the growing season.

- **Scab (*Cladosporium carpophilum*).** A fungus causing spotting and blotching on leaves, fruit and twigs; it can result in premature leaf fall. Occurs in most regions. Favoured by wet weather, but not usually serious enough for action to be taken, although a copper fungicide can be used in spring (2-5 weeks after petal fall). Cut out diseased shoots in the growing season.

- **Hull rots.** Several moulds can attack the fruit and fruiting wood, including *Rhizopus stolonifer*, *Monilinia fructicola* and *M. laxa* (brown rot – see also 'Brown rot blossom blight' left). These occur in most regions, most severely on vigorous-growing, soft-shelled American cultivars. Lower leaves and spurs are killed a few weeks before harvest, and hulls are attacked both outside and (if split) on the inside. Minimize damage by harvesting as soon as hulls split, and by reduce nitrogen input to reducing vigour. Make sure no diseased nuts are left on the tree after harvest.

Harvesting and yields

The traditional harvest method is to shake trees or individual branches over canvas sheets. Mechanical harvesting is now standard practice in Californian and other large commercial orchards, using shaking machines and pick-up equipment (necessitating level orchards). Hand-harvesting using Nut Wizards (hazelnut size) works well.

All parts of the almond fruit reach their final size months before harvest; in the kernel, however, dry matter continues to accumulate. The first sign that the nut is maturing is an indented 'V' followed by a split along the suture of the hull. Harvesting is usually 30-45 days after this point, during which time the split hull continues to open, exposing the nut inside. A separation (abscission) zone forms between the fruit and the stalk, and when fully formed the nut is attached to the tree only by a few remaining fibres: these are broken during harvest shaking or wind, and the nut falls from the tree.

Time of harvesting is late summer and autumn (mid- to late autumn in cool climates), depending on cultivars and climate. To determine when to harvest, shake a branch with your hands and if nearly all nuts fall, then harvest immediately (as long as the weather is dry). The nuts in the centre of the tree are last to ripen, and nuts ripen earlier on non-irrigated trees than on irrigated trees. Make sure all nuts are removed from the tree, because those left on can harbour fungal diseases and pests like the navel orangeworm.

Yields for a particular variety vary from year to year and are dependent on a host of factors, including location, pollenizing conditions, orchard management, etc. Cropping usually begins by about the fourth year after planting (depending partly on the rootstock used). The approximate ranges given below vary from non-fertilized, non-irrigated trees to highly irrigated and fertilized commercial orchards; yields here are those of nut kernels. It is common for almonds to crop biennially, i.e. with a light crop one year followed by a heavy crop the next, and so on.

Processing and storage

After harvesting and hull removal, nuts for drying are dried to a nut moisture of 7 per cent. Dried in-shell nuts can be stored for up to 20

Years from planting	Yield per tree	Yield per hectare	Yield per acre
4-6	0.5-2.5kg (1-6lb)	92-470kg (203-1,036lb)	37-188kg (82-414lb)
7-12	3.5-18kg (8-40lb)	650-3,375kg (1,433-7,440lb)	260-1,350kg (573-2,976lb)
12-15	5-18kg (11-40lb)	937-3,375kg (2,065-7,440lb)	375-1,350kg (827-2,976lb)

Almond cultivars 'Ferraduel', 'Ferragnes' and 'Robijn'. 'Robijn' is a hybrid and has rougher shells.

months at 0°C (32°F) or 16 months at 10°C (50°F) or 8 months at 20°C (68°F). Commercial in-shell nuts are usually bleached with sulphur dioxide to make their appearance more attractive.

Commercial considerations

* Crops are more weather-dependent at flowering time than most nuts.
* Competing with commercial growers to sell in-shell nuts is very difficult.
* More commercial potential after processing nuts into other forms.

Propagation

Almond cultivars are either budded or grafted on to a suitable rootstock; any of the common methods is applicable.

Cultivars

Although most of the cultivars listed here are true almond, the species has also been hybridized with peach to create cultivars ('Ingrid', 'Robijn') that have almond-like leaves and fruit, and sweet kernels borne inside nuts that look more like peach stones – they have a rougher exterior than true almond nuts. Hybrids are hardier and can have resistance against peach leaf curl.

In cool temperate climates, the most promising cultivars are those that flower late but mature early season. Both older and more recent Dutch and French varieties show the most promise.

The kernel percentage (i.e. what percentage the kernel is of the whole nut) varies (mainly 25-45 per cent), with hard-shelled cultivars at the lower end of the scale and soft-shelled (thin-

shelled) higher. Very thin-shelled cultivars are sometimes called paper-shelled.

Shell softness or hardness varies and is related to the shelling yield (or percentage of kernel within the whole nut). Classifications are:

Very hard-shelled	= 20-25% kernel
Hard-shelled	= 25-35% kernel
Semi-hard-shelled	= 35-45% kernel
Soft-shelled	= 45-55% kernel
Paper-shelled	= 55-65% kernel

Kernel size varies between cultivars:

* **Small** kernels yield approximately 400-450 kernels/kg (880-1,000/lb).
* **Medium-size** kernels 350-400 kernels/kg (775-880/lb).
* **Large** kernels 270-350 kernels/kg (600-775/lb).

The kernel colour also varies from light to dark.

There are hundreds of almond cultivars and it is impossible to list them all here. However, all the main commercial and home-grown cultivars are listed below.

Key to almond cultivar table
Flowering

Some almond cultivars fall into specific flowering groups within which all are related. These are called pollen incompatibility groups. Cultivars in the same group will not pollinate each other. If a flower group number is given in the description below, that cultivar will not pollinate others with the same group number.

The flowering time is divided into the following categories. Each of these lasts about a week, so the total flowering period varies by around six weeks. 'E' roughly corresponds to late winter, and 'VL' to early spring or later. Cultivars in the same or an adjacent group either side will cross-pollinate, unless they are in the same incompatibility group:

E = early
EM = early–mid
M = mid
ML = mid–late
L = late
VL = very late

Harvesting

Almond ripening can vary over a period of about 60 days (2 months):

E (early) is 0-7 days
EM (early–mid) is 7-15 days
M (mid) is 15-25 days
ML (mid–late) is 25-40 days
L (late) is 40-60 days

The dates corresponding to the 2-month period vary with climate. In cool climates it is mid–late autumn. In warmer climes it can be late summer and early autumn.

Recommended almond cultivars

Cultivar	Origin	Flowering period						Harvesting period					Description
		E	EM	M	ML	L	VL	E	EM	M	ML	L	
'Ai'	France						✓			✓			Kernels 40% of nut. Tree bushy, a regular producer, resistant to *Monilinia*, hard to prune.
'Aldrich'	California			✓						✓			Nuts soft-shelled, well sealed, kernels small to medium. Tree large, fairly upright.
'All-in-One'	California					✓		✓					Nuts well sealed, soft-shelled, medium-size kernels. Tree semi-dwarf, heavy cropping, precocious (early flowering), self-fertile. Bred for home gardeners.
'Antoñeta'	Spain					✓					✓		Nuts hard-shelled; kernels large. Tree self-fertile, spreading, vigorous.
'Ardechoise'	France		✓							✓			Nuts soft-shelled, long kernels. Very upright tree, productive, disease-resistant.
'Ayles'	Spain					✓					✓		Nuts hard-shelled, 30-35% kernel; kernel heart-shaped, medium–large, good quality. Tree self-fertile, moderately vigorous, spreading, compact growth, tolerant to late frosts.
'Belle d'aurons'	France						✓						Nuts large, flat, 35% kernels; kernels good quality. Tree large, very resistant to *Monilinia*.
'Belona'	Spain					✓				✓			Nuts hard-shelled; large kernels of good quality. Tree self-fertile, semi-upright, moderate vigour.
'Blanquerna'	Spain			✓				✓					Nuts hard-shelled, good-quality kernels. Tree open, self-fertile.
'Butte'	California				✓						✓		Semi-hard-shelled nuts, kernels small. Tree spreading, moderate vigour, productive, susceptible to brown rot. Flower group 7.
'Cambra'	Spain					✓				✓			Nuts hard-shelled; kernels good quality. Tree self-fertile, open, moderate vigour.
'Carmel'	California				✓						✓		Thin soft-shelled nuts, kernels small, long, pale. Tree upright, moderate vigour, precocious, productive. Susceptible to brown rot. Flower group 5.

Recommended almond cultivars

Cultivar	Origin	Flowering period						Harvesting period					Description
		E	EM	M	ML	L	VL	E	EM	M	ML	L	
'Chellaston'	Australia	✓								✓			Nuts soft-shelled; kernels flattish, small–medium size. Tree semi-upright, compact, moderate vigour.
'Constanti'	Spain				✓					✓			Nuts hard-shelled; large kernels. Tree self-fertile, semi-upright.
'Cruz'	California			✓						✓			Nuts soft-shelled, well-sealed; kernels round, medium size. Tree upright, open, moderate vigour, a heavy consistent producer.
'Davey'	California							✓					Nuts soft-shelled; medium-size kernels, pale. Tree vigorous, upright, slow to bear, low productivity, difficult to harvest.
'Desmayo Largueta'	Spain											✓	Nuts hard-shelled, pointed. Tree precocious. One of the main Spanish cultivars.
'Falsa Barese'	Italy				✓						✓		Nuts fairly hard-shelled; kernels 33-39%. Tree productive.
'Felisia'	Spain						✓	✓					Nuts hard-shelled; small kernels. Tree open, self-fertile, moderate vigour.
'Ferraduel'	France						✓			✓			Nuts hard-shelled, 28% kernels; kernels large, flat. Tree vigorous, precocious, productive, resistant to peach leaf curl. An important cultivar in new European plantings – used to pollinate 'Ferragnes'.
'Ferragnes'	France						✓			✓			Nuts hard-shelled, 30-43% kernels; kernels large, elongated, light-coloured, somewhat wrinkled. Tree moderately vigorous, upright, precocious, productive, resistant to peach leaf curl and *Monilinia*. An important cultivar in new European plantings – used to pollinate 'Ferraduel'. Flower group 8.
'Ferralise'	France						✓	✓					Nuts hard-shelled, 30% kernel; kernels small, elongated, smooth. Flower group 8.
'Ferrastar'	France						✓			✓			Tree vigorous, resistant to some fungus diseases.
'Filippo Ceo'	Italy				✓			✓					Nuts soft-shelled; kernels 40%. Used for almond flour and paste.

Recommended almond cultivars

Cultivar	Origin	Flowering period						Harvesting period					Description
		E	EM	M	ML	L	VL	E	EM	M	ML	L	
'Francoli'	Spain					✓				✓			Nuts hard-shelled; kernels medium size, 30%. Tree vigorous, precocious, self-fertile.
'Fritz'	California			✓								✓	Nuts semi-hard-shelled; kernels fairly small. Tree upright, vigorous, heavy cropping.
'Garden Prince'	California			✓									Nuts soft-shelled; kernels medium size. Tree self-fertile, a genetic dwarf to 3m (10') high. Bred for home gardeners.
'Gaura'	Spain					✓		✓					Nuts very hard-shelled, 30-35% kernel; kernel medium–large, good quality. Tree self-fertile, productive, frost-tolerant, spreading, medium size.
'Glorieta'	Spain					✓				✓			Nuts hard-shelled; kernels large, 29%. Tree very vigorous, densely branched, precocious.
'Hashem II'	California			✓	✓				✓				Nuts soft-shelled; kernels large, long, flat. Tree upright and productive.
'Ingrid'	Netherlands						✓			✓			A peach–almond hybrid with pink flowers and thick-shelled nuts of good flavour. Tree quite resistant to peach leaf curl.
'IXL'	USA	✓							✓				Nuts long with pronounced keel, shell soft to paper; kernels medium–large, pale. Tree upright, large leaves, light bearer. Flower group 1.
'Jordanolo'	California	✓								✓			Nuts soft-shelled; kernels pale, large, good quality. Tree upright, productive, precocious.
'Kapareil'	California			✓					✓				Nuts paper-shelled; small kernels. Otherwise similar to 'Nonpareil'. Flower group 6.
'Lauranne'	France					✓			✓				Hard-shelled, 38% kernel; kernel small, pale. Tree spreading to drooping, moderate vigour, precocious, self-fertile.
'Le Grand'	California				✓					✓	✓		Tree partly self-fertile, vigorous, upright, susceptible to brown rot and shot hole.

Recommended almond cultivars

Cultivar	Origin	Flowering period						Harvesting period					Description
		E	EM	M	ML	L	VL	E	EM	M	ML	L	
'Livingston'	California					✓					✓		Nuts thin-shelled, well sealed; medium-size kernels. Tree semi-upright, moderate vigour. Flower group 5.
'Lodi'	California			✓					✓				Nuts soft-shelled, well sealed; medium–large broad kernels are slightly bitter. Tree moderate vigour.
'Macrocarpa'	UK	✓									✓		Nuts large. Tree with large flowers, resistant to peach leaf curl, long grown in the UK.
'Mandaline'	France					✓			✓				Kernels of very good quality. Tree self-fertile and a good pollinator, resistant to fungal diseases.
'Marcona'	Spain			✓								✓	Nuts hard-shelled; kernels large, round, 25-28%. One of the main Spanish cultivars.
'Mardia'	Spain						✓	✓					Nuts hard-shelled; good-quality kernels. Tree self-fertile, semi-upright, vigorous, disease-resistant.
'Marinada'	Spain						✓		✓				Nuts hard-shelled; large kernels. Tree self-fertile, semi-upright, moderate vigour.
'Marta'	Spain				✓				✓				Nuts hard-shelled; kernels large. Tree self-fertile, upright, vigorous.
'Masbovera'	Spain				✓				✓				Nuts hard-shelled; kernels 28%, large. Tree very vigorous, densely branched.
'Merced'	California			✓							✓		Nuts paper-shelled; medium-size pale kernels. Tree upright, small–medium size, precocious, susceptible to worm damage and fungi. Flower group 3.
'Mission' ('Texas')	USA (Texas)				✓							✓	Nuts semi-hard-shelled; medium-size kernels, dark, many doubles. Tree upright, very productive, vigorous when young.
'Monarch'	California			✓					✓				Nuts quite hard-shelled, well sealed; kernels large and plump. Tree large, upright. Flower group 5.

Recommended almond cultivars

Cultivar	Origin	Flowering period						Harvesting period					Description
		E	EM	M	ML	L	VL	E	EM	M	ML	L	
'Moncayo'	Spain						✓				✓		Nuts very hard-shelled, 25-28% kernel; kernel medium–large, good quality. Tree vigorous, medium–large, spreading to drooping, self-fertile, tolerant of late frosts.
'Monterey'	California				✓							✓	Large, elongated kernels. Tree vigorous, spreading, very productive. Flower group 7.
'Ne Plus Ultra'	California	✓								✓			Nuts paper-shelled; kernels very large, pale, many doubles. Tree vigorous, spreading, precocious, a moderate cropper, difficult to train. Susceptible to frost, fungus and worm damage. Flower group 3.
'Nonpareil'	California			✓				✓					Nuts paper-shelled; medium-size pale kernels. Tree large, upright-spreading, vigorous, a consistent heavy cropper. Fairly resistant to frost damage. The leading Californian cultivar. Flower group 1.
'Padre'	California					✓					✓		Nuts hard-shelled; small kernels. Tree upright, productive, moderate vigour.
'Peerless'	California		✓					✓					Nuts semi-hard-shelled; kernels pale, mediocre quality – used for in-shell nut sales. Tree a moderate cropper with moderate vigour.
'Penta'	Spain						✓	✓					Nuts hard-shelled. Tree self-fertile.
'Phoebe'	Netherlands				✓								Flowers late, with pink ornamental blooms. Tree self-fertile, resistant to peach leaf curl.
'Plateau'	California			✓						✓			Nuts soft-shelled, well sealed; large kernels. Tree semi-upright, moderate vigour.
'Price'	California			✓				✓					Nuts soft-shelled; good-quality kernels. Tree a heavy biennial cropper. Flower group 3.
'Rabasse'	France						✓						Nuts small, round, hard-shelled; 30% kernel. Tree highly productive.
'Robijn'	Netherlands						✓			✓			Nuts with a soft shell, resistant to peach leaf curl. A peach–almond hybrid.

Recommended almond cultivars

Cultivar	Origin	E	EM	M	ML	L	VL	E	EM	M	ML	L	Description
		Flowering period						**Harvesting period**					
'Ruby'	California					✓							Nuts hard-shelled. Tree small, upright, productive.
'Shefa'	Israel	✓						✓					Nuts soft-shelled; good-quality kernels. Tree precocious, vigorous, productive.
'Solano'	California			✓					✓				Similar to 'Nonpareil' with high-quality kernels. Flower group 6.
'Soleta'	Spain					✓					✓		Nuts hard-shelled; kernels large, excellent roasted. Tree semi-upright, moderate vigour, self-fertile.
'Sonora'	California		✓						✓				Nuts thin-shelled; kernel pale, large, smooth, long. Tree rounded, precocious, heavy bearing, moderate vigour, frost-resistant flowers. Flower group 6.
'Steliette'	France					✓	✓	✓					Nut semi-hard-shelled; 45% kernel; kernel large, pale. Tree precocious, moderately vigorous, self-fertile.
'Tardona'	Spain						✓			✓			Nuts hard-shelled; kernels small. Tree self-fertile, densely branched.
'Tarraco'	Spain						✓			✓			Nuts hard-shelled; kernels very large. Tree semi-upright, moderate vigour, disease-tolerant.
'Thompson'	California					✓				✓			Nuts soft- to paper-shelled; kernels plump, small, slightly bitter. Tree upright, precocious, productive, moderate vigour. Flower group 4.
'Titan'	USA					✓							Nuts thin-shelled, well sealed. Tree hardy, large, resistant to peach leaf curl.
'Tuono'	Italy						✓	✓					Nuts hard-shelled; kernels pale, large, many doubles. Tree self-fertile, spreading, productive, moderate vigour, low susceptibility to pests and diseases.
'Vairo'	Spain			✓				✓					Nuts hard-shelled; kernels large. Tree self-fertile, vigorous.
'Wood Colony'	California		✓						✓				Nuts quite soft-shelled, well sealed; medium-size plump kernels. Flower group 4.

BLACK WALNUT

(Juglans nigra)

ZONE 4, H7

The black walnut is native to eastern North America – eastern USA as far north as the Canadian border (hence its alternative common names of Virginian walnut, American walnut, eastern black walnut) – and has been cultivated for a long time in Europe, where it is now naturalized. It is often grown as a timber tree for the fast-growing, high-quality timber.

A young black walnut tree.

It is a large, fast-growing deciduous tree, growing up to 50m (160') high in its native habitat, though in cool climates rarely more than half that. Pyramidal when young, it becomes spreading and round-crowned with age, though usually with a long trunk. It has brownish-black bark, deeply furrowed into diamond-shaped ridges, and downy aromatic young branches (an easy way to tell it apart from the common or English walnut, *Juglans regia*). It casts quite a dark shade. Typical lifespan is about 150 years, though some trees can live to 300 years of age or more.

The compound leaves are 30-60cm (1-2') long with 15-23 leaflets each about 6-12cm (2½-5") long; they are fragrant when rubbed. The leaflets are a glossy dark green above and downy beneath; the terminal leaflet is often small or absent – another good identifying feature. Leaves turn a bright yellowish-gold in autumn.

Male catkins are 5-10cm (2-4") long, developing from the leaf axils of the previous year's growth. Tiny female flowers occur in terminal spikes of 2-5 small green flowers borne on the current year's shoots. Flowering and fruiting of seedling trees begins at about 12-15 years of age, and grafted trees at 6-8 years. Black walnuts are wind-pollinated.

Fruit are borne in ones or twos, being round, 4-5cm (2") wide, with a thick green hull (husk) enclosing a single nut, which is irregularly and longitudinally furrowed, with rough edges. The husks turn from green to yellowish-green when ripe, and usually drop intact with the nut inside. The nuts are 25-40mm (1-1⅝") across, sometimes more, thick-shelled and enclose an edible kernel. Fruiting is often biennial, with heavy crops every other year.

The root system usually consists of a number of deep taproots, which can penetrate more than 2m (6'6") deep, with long lateral roots and feeder roots that normally concentrate at a depth of 10-20cm (4-8").

Uses

The nut kernels are edible and excellent, with a more robust, fuller and richer flavour than English walnuts. The flavour is retained on baking, hence many of the traditional American recipes using it are for baked foods including cakes, pies, breads, etc.; ice cream is another traditional use. The kernels are high in polyunsaturated fats, protein and carbohydrates, plus Vitamins A, B, C and linoleic acid (see Appendix 1). They store well for only 6-12 months. A drawback is that black walnuts are one of the hardest nuts to crack, and require specialized heavy-duty crackers, which are available in North America.

An edible and delicious oil is expressed from the kernels and used raw or cooked. Like the nuts, it doesn't store for very long.

The unripe fruit (picked in July) are pickled in vinegar (husk and all) – similar to European walnuts.

Secondary uses of black walnut

The sap of the tree is edible. It can be tapped in the same way as maple sap from maples, and can be consumed fresh, concentrated to make a syrup, or used to make wine, beer, etc.

The shells left over from removing kernels

can be ground up and are then used commercially as an excellent abrasive (very hard, light, non-toxic, doesn't pit or scar). Used on stone, metals and plastics, and also as the gritty agent in some soap and dental cleansers (they were even used by NASA to clean the exterior surfaces of the space shuttle); they are also used in paints, glue, wood cements, oil and well-drilling, sand-blasting machines, in tyres for added winter traction on ice and as a filler in dynamite.

The green husks (hulls) left over from husking machines in commercial cultivation are a valuable pasture fertilizer. They are high in nitrogen and phosphorus, and (though they contain anti-germinant chemicals, which can be detrimental to annual crops) perennial grasses and clovers thrive with a husk mulch; earthworm populations are also stimulated. Each kilogram of husked walnuts yields about 2kg of husks, hence large quantities of husks can soon be generated. The husks don't compost well (they are too heavy and a pile becomes anaerobic) and are best applied fresh to pasture. Recommendations from organic farmers in North America who use husks successfully are to apply at 25-38t/ha or 2.5-3.8kg/m² (10-15t/acre); a drawback about using it fresh in autumn is that husk breakdown may not be complete when winter temperatures stop grass growth, and leaching of the nutrients may then occur. However, spreading fermenting husks can result in chemical imbalance problems in the topsoil. Growth stimulation of the grasses begins within a short time of application – weeks rather than months.

The bark, husks and leaves have all been used in traditional North American medicine. These parts contain juglone, known to be antihaemorrhagic (used to stop bleeding) and fungicidal/vermifugal (the leaves and husks are used against skin fungi like athlete's foot and parasites like ringworm). An extract from the heartwood is used in treating equine laminitis.

Good colour-fast dyes are obtained from the husks, leaves and bark. The husks readily stain the skin with a persistent brown stain, and have long been used to dye wood, hair, wool, linen and cotton. The bark and fresh green husks dye yellowish-brown with an alum mordant and brown with an alum mordant; the dried husks dye golden-brown (alum mordant), dark brass (chrome mordant), coffee (copper mordant), camel (tin mordant), charcoal grey (iron mordant) and light brown (no mordant).

In North America the black walnut has been widely used as a rootstock for the common walnut (*Juglans regia*) in the past, but this is

Black walnut fruit.

no longer so common because of 'blackline disease', which causes a delayed failure of the graft union when the rootstock is a different species of walnut.

Black walnut is highly valued as a timber tree in many areas, including North America, and Austria, France, Germany, Hungary, Romania and former Yugoslavia in Europe; it is seen as a high-quality replacement for diminishing tropical hardwoods. It is usually grown on a fairly short 30-50 years' rotation. American studies comparing the costs and returns of black walnut and Douglas fir plantations show the walnut to be about 7 times as profitable over an 80-year rotation.

The timber is coarse and mostly straight-grained, a rich dark brown to purplish-black with light sapwood. Strong, tough, extremely durable, heavy and hard, it is easy to work and resistant to fungi and insect pests. Black walnut is valued for high-quality cabinet work, joinery, shipbuilding, musical instruments, veneers, gunstocks and plywood; it also makes excellent fuel.

Agroforestry uses of the walnut family

Trees of the walnut family all grow large in time and are therefore usually planted at wide spacing for nut production. The space between trees in the first 10-20 years leads to various intercropping options (see Chapter 1, page 34).

The leaves of all walnut species contain a substance, juglone, which has growth-suppressing allelopathic effects on many other plants, including apples; this occurs beneath and near the walnut canopy where leachate from rain falling on leaves, leaf litter, and walnut root exudates can affect other plants. In practice, plants in alleys or gaps between walnuts are unlikely to be affected for some years – basically until the walnut roots and intercrop roots start to meet and mingle, which may be 10 years or more for plants in the centre of wide alleys. Juglone is rapidly degraded in the soil by bacteria, so that root-to-root contact is more likely to cause negative effects than juglone from leaf litter or leachate.

Walnuts can be intercropped with pasture, for cutting hay, market gardening crops, arable crops, Christmas trees, nurse trees / short-rotation tree crops, or nitrogen-fixing shrubs.

Pasture in alleys can be used for grazing, once trees are large enough.

Horticultural crops under consideration for intercropping in the USA include vegetables, soft fruit, bare-rooted nursery stock and flower bulbs. Crops can be grown for 9-15 years before shading effects become large.

Several arable crops have been intercropped with black walnuts, including sweet corn, soya beans and wheat. The advantage of winter wheat is that it grows during the walnuts' dormant season. The alleys used for the arable crop must be set to a convenient width for tractor cultivations, etc.; and these alleys must be slowly reduced in width year by year, as shading and juglone effects gradually increase. A grass- and intercrop-free strip each side of the walnuts must be maintained, which needs to be between 1m (3') for winter wheat and 2m (6-7') for spring-sown intercrops. With rows of walnuts planted 12m (40') apart, arable crops can grow for 9-15 years before shading becomes a problem.

Douglas firs grown as Christmas trees have been successfully intercropped with black walnut in Oregon: black walnuts planted at 4.5m (14'6") spacing, with the firs at 1.5m (5') spacing between. The firs are harvested by 7 years after planting – longer than this and they start to show signs of juglone growth inhibition. The firs also act as a nurse crop to the walnuts and encourage straighter growth.

Apart from Christmas trees, other nurse trees can be interplanted to aid the early growth of the walnuts and force straighter growth. Some of the nitrogen-fixing trees can achieve this, for example alder (*Alnus glutinosa*) and black locust (*Robinia pseudoacacia*), but care must be taken over species choice so that the walnuts are not outcompeted too quickly, and that when the intercrop is removed, the stumps remaining are not susceptible to honey fungus (*Armillaria* spp.). Alder is recommended in North America, and is removed after 12-15 years (by which time it is being affected by juglone) for firewood.

There is plentiful evidence from North America that interplanting walnut plantations with alders or *Elaeagnus* improves the growth of the walnuts substantially (improvements in diameter and height of 20-50 per cent plus). Autumn olive (*E. umbellata*) has

performed particularly well in this respect, and is recommended as shrub interplant; oleaster (*E. angustifolia*) and Siberian pea tree (*Caragana arborescens*) also succeed – best in continental climates. When using autumn olive, note that it is faster growing than the walnuts, so either plant 4-5m (13-16') away from walnut plants, or plant after the walnuts have established and are already above head height.

Cultivation

Siting requirements are a sheltered sunny site, which is not susceptible to late spring frosts. An ideal would be mid-slope on a sheltered south or south-west aspect.

Planting should be at a spacing of 8-15m (26-50') apart. Although trees can grow into huge, wide-spreading specimens with crowns 15m

Black walnut fruits after falling from the tree.

(50') across, trees planted at closer spacings will not be at risk from crowding for many years and pruning/thinning can always be carried out if necessary.

Weed control is essential for 2-4 years, competition from grasses being the most common cause of young tree deaths. Mulching is best to exclude weeds. Weed control to a radius of 2m (6'6") is desirable for several years.

Rootstocks and soils

Trees should be grown on their own roots or grafted on to black walnut or common walnut rootstocks.

Soil requirements are a moderately fertile, deep, well-drained soil of medium texture and near neutral pH (6 to 7). Very sandy and clayey soils are unsuitable, though growth is good on chalk and limestone, where there is at least 60cm (2') depth of soil. Because they are deep-rooting, trees are very drought-resistant once established.

Pollination

Black walnuts bear male and female flowers on the same tree, but they usually mature at different times, with the female flowers most often preceding the males by about five days. Because of this, self-pollination is unlikely and trees/cultivars should be assumed not to be self-fertile. Pollination is via the wind.

Male flowers develop from leaf axils towards the ends of the previous year's branches; female flowers originate from terminal buds on the current year's shoot and appear before the leaves are fully expanded in spring. Depending

Male catkins of black walnut.

on the climate and selection, the flowers appear between mid-spring and midsummer. Although black walnuts will cross-pollinate with Persian/English walnuts, they tend to flower some 2-3 weeks later, hence pollination is unlikely, except with early flowering black walnuts and late flowering Persian walnuts.

Flowering occurs over a period of about 10 days. It is wise to choose two or more cultivars that will cross-pollinate (see the cultivar table below).

Feeding and irrigation

Like most fruiting trees, the main nutrient requirements are for nitrogen and potassium. Too much feeding, though, can make the tree susceptible to fungal diseases. If possible, mulch with manures or compost and/or use nitrogen-fixing plants and potassium accumulators like comfrey nearby to top up nutrient levels. Irrigation is unnecessary.

Pruning

Pruning is generally unnecessary, although a little formation pruning may be desirable to form a tree of particular shape.

Pests

The insect pests of significance in North America are walnut husk flies (*Rhagoletis completa*), which feed on the green husk of nuts, producing a staining and off-flavouring of the kernel; curculio (*Conotrachelus retentus*), which feed on young leaves and husks – control by collecting and destroying prematurely fallen nuts; and fall webworm (*Hyphantria cunea*, a moth larva that feeds on foliage). There are no significant insect pests in Europe.

As for wildlife pests, the fleshy and strong-smelling husks deter squirrels (to an extent) and other pests from eating the nuts; but once the nuts are de-husked (whether naturally or manually), squirrels are extremely fond of the nuts and will take them to bury for the winter if allowed. Rabbits and deer can occasionally browse foliage. Crows can occasionally be a serious pest, attacking nuts on the trees.

Diseases

Black walnut can share diseases with European walnut. It is rarely troubled by walnut blight, but can sometimes suffer from walnut leaf blotch.

Walnut leaf blotch / anthracnose (*Gnomonia leptostyla*) is widespread in North America and Europe. The fungus causes brown blotches on leaves, and can cause defoliation and infection of the developing fruit, which then drops; less severe infections can reduce kernel weights or darken the kernels. The disease appears in late spring, favoured by wet weather, and its spores overwinter on dead leaves. One control, if attacks are consistently bad, is to collect fallen leaves and compost at high temperatures or burn. In wet seasons when infection is bad, copper-based sprays such as Bordeaux mixture give effective control. Most cultivars are resistant when young and several continue to be resistant when older.

Harvesting and yields

Black walnuts are often irregular bearers, fruiting biennially or in between annual and biennial cropping. Nuts from black walnuts have very hard, dark brown or black shells with irregular grooves and ridges; the kernels are stronger flavoured than common walnuts. The husks turn slightly yellow when the nuts are ripe and ready to fall.

Typical yields of grafted cultivars are about 8kg (18lb) of nuts (around 350 nuts) per year for 15-year-old trees, rising to a maximum at 50-60 years, when yields can reach 100kg (220lb) or more; 20-year-old plantations can yield 5t/ha (2t/acre). These are yields in shell; kernel yields are about 30 per cent of these. Trees are long-lived and can continue to crop for 90 years or more.

The nuts usually reach full size in late summer and ripen in mid- or late autumn; they drop, usually within the green fleshy husks, just before the leaves fall. At this stage, they are quite easily shaken from the tree. Ripeness can be checked by feeling a husk – these soften as ripeness approaches.

After harvest, the husks must be removed within a few days and the nuts washed, as the husks darken rapidly and can affect the kernel colour and flavour if left attached too long. Wear gloves, as handling the green husks can leave stains on hands and clothes, and are very difficult to remove. Loosen and remove husks by rolling between hands, stomping with feet, by using a cement mixer with a brick or two in, or by driving over nuts in a vehicle (really!). During washing, the bad nuts can be separated out, as they will float – all good nuts will sink.

Processing and storage

Dry nuts at temperatures of 30-40°C (86-104°F). Storage of dried nuts is best with whole nuts, rather than kernels only. Low temperatures near zero are preferable, with low humidity. Dry nuts can be stored for up to a year, but eventually they will turn rancid.

Before shelling or cracking nuts, they should be soaked in water overnight to moisten them and strengthen the kernels, otherwise the shells shatter badly and kernels may break in the cracking process. There are several specially designed hand and mechanical crackers in North America to cope with black walnuts, though a carefully controlled hammer can do the job! The ordinary type of nutcracker will not cope, probably breaking itself not the nut.

Propagation

On average there are about 90 nuts/kg (41 nuts/lb). Seeds require stratification for about 16 weeks before sowing: mix with moist sand or compost and keep cold (in a fridge, for example); keep an eye out for roots starting to emerge in the spring.

Seeds should be sown in deep containers (e.g. 'Rootrainers') or seedbeds and covered with 25-50mm (1-2") of media. Predation from rodents can be a bad problem, especially with outside seedbeds; these should be protected, and deep containers should be kept off the ground.

Germination occurs within about 3-5 weeks. The average germination rate is about 50 per cent, and seedlings grow rapidly to a height of 30-60cm (1-2') in the first year.

The cultivars 'Beck', 'Fonthill', 'Minnesota Native', 'Myers' (syn. 'Elmer Myers'), 'Patterson', 'Putney' and 'Thomas' are noted for their vigour and straight form; seed from these is likely to produce a higher percentage of timber trees of good form than unnamed seedlings. Late leafing is also highly heritable; from the above list, 'Myers' and 'Thomas' are very late leafing.

Grafting is difficult and requires the use of a hot graft pipe (see Chapter 2, page 61). Established trees can be top-worked outside in late spring. Early summer budding or greenwood (semi-ripe) grafting have the best chance of success.

Cultivars

More than 100 named cultivars have been selected and many are still grown in North America. The best known are 'Wiard' (Michigan); 'Cochrane' and 'Huber' (Minnesota); 'Thomas' (New York); 'Cornell' and 'Snyder' (northern areas); 'Ohio' in central USA; 'Myers', 'Sparrow' and 'Stambaugh' (southern USA). Many have been selected for their cracking quality, including the so-called peanut types. These are single-lobed sports that have only half a kernel, which can be extracted whole (e.g. 'Blaettner', 'Throp', 'Worthington').

Cultivars differ in hardiness, response to climatic conditions, resistance to diseases, and susceptibility to insect damage. There is considerable variation in nut quality, flowering and leafing dates, precocity (age of bearing), and growth rate. Most bear their fruit at the tips of branches, though some selections that fruit on lateral buds have been reported. 'Myers' and 'Victoria' are reported to be partly self-fertile.

Because of the large shell, the actual percentage of the nut that forms the kernel is relatively low – 30 per cent is good (36 per cent is very good, 27 per cent the average).

Although using grafted cultivars is the most reliable method of producing nuts, seedlings from good cultivars tend to reproduce the parent tree's tree and nut characteristics. Seedling trees are of value and maybe lower cost.

The cultivars rated most highly on a range of attributes by American nut growers include 'Emma K', 'Hay', 'Myers', 'Ohio', 'Rowher', 'Sparrow' and the 'Sparks' selections. Selections that ripen their nuts well in cooler climates include 'Beck', 'Bicentennial', 'Bowser', 'Davidson', 'Emma K', 'Grimo 108H', 'Hare', 'Krause', 'Myers', 'Ohio', 'Pfister', 'Sparks 127', 'Thomas' and 'Weschcke'.

Black walnut Emma K

Black walnut Thomas

Black walnut Weschcke

As with other walnut species, in most trees the male and female flowers are mature at different times. Although there is a lack of information for most varieties, the following list identifies group A trees, which have male flowers preced-ing females, and group B, which have females preceding males. A mix of group A and B will ensure good pollination. Seedling trees will be more variable and a mixture of several should give satisfactory pollination.

Recommended black walnut cultivars

Cultivar	Origin	Description
'Baker's	Ohio	Annual bearer. Thin shells, higher kernel percentage.
'Baum #25'	Kentucky	Mid-season, moderate biennial bearer. Nuts crack well, thin shell.
'Beck'	Michigan	Early leafing, self-fertile, early season. Medium-size nuts with thin shells and large kernels. Vigorous, heavy cropper.
'Bicentennial'	New York	Large nuts and kernels. Cracks well.
'Bowser'	Ohio	Mid-late leafing, group B. Medium-size nuts with thin shells and large kernels. Vigorous, heavy bearer.
'Burns'	Ontario	Thin-shelled nuts that crack well.
'Burton'	Kentucky	Upright tree. Thick-shelled nuts.
'Clermont'	Ohio	Very late leafing, late season. Large nuts, thin-shelled, with large kernels. Heavy bearer.
'Cornell'	New York	Early season. Medium-size nuts with large kernels.
'Cranz'	Pennsylvania	Mid-late leafing, group A, late season. Medium-size nuts with large kernels. Lateral bearer.
'Davidson'	Iowa	Early leafing, group A, early season. Lateral bearing.
'Drake'	Unknown	Thin-shelled. Heavy bearer.
'Edras'	Iowa	Thin-shelled nuts with large kernels.
'El-Tom'	Unknown	Thin-shelled nuts with large kernels.
'Emma K'	Illinois	Medium-size nuts with large kernels, thin shells. Heavy and regular bearer.
'Evans'	Unknown	Vigorous upright tree. Very large nuts.
'Farrington'	Kentucky	Late leafing, group B, early season. Large nuts with large kernels, thin-shelled. Regular heavy bearer.
'Fonthill'	Unknown	Medium nuts. Vigorous upright heavy bearer.
'Football II'	Missouri	Very large nuts. Lateral bearing.
'Grimo 108H'	Ontario	Large nuts with large kernels.
'Grundy'	Iowa	Early leafing, group A. Large nuts with large kernels.

Recommended black walnut cultivars

Cultivar	Origin	Description
'Hain'	Michigan	Mid-season. Peanut type with thin shells.
'Hare'	Illinois	Mid-leafing, group B. Large nuts with large kernels.
'Harney'	Kentucky	Thin-shelled nuts with large kernels.
'Hay'	Missouri	Mid-leafing, group B.
'Homeland'	Virginia	Early season. Large kernels.
'Krause'	Iowa	Early leafing, group A, late season. Very large nuts, moderate kernels. Good pollinator.
'Lamb'	Michigan	Vigorous upright tree. Nuts medium-large.
'Majestic'	Uncertain origin	Large kernels. Reliable bearer.
'Mintle'	Iowa	Medium-sized thick shells. Heavy cropper.
'Monterey'	Pennsylvania	Large nuts and kernels.
'Myers' (syn. 'Elmer Myers')	Iowa	Late leafing. Self-fertile, late season. Medium-size nuts with large kernels.
'Ogden'	Kentucky	Early–mid-leafing, self-fertile. Large nuts, thin shells, moderate kernels.
'Ohio'	Ohio	Early–mid-leafing, group B. Medium-size, thick-shelled nuts with moderate kernels. Biennial bearer.
'Patterson'	Iowa	Vigorous upright tree. Large nuts.
'Peanut'	Ohio	Mid-leafing, group B, early season. Some nuts single-lobed.
'Pfister'	Nebraska	Mid-leafing, group A, early season. Large kernels. Lateral bearing.
'Pinecrest'	Pennsylvania	Large nuts and kernels.
'Putney'	New York	Early season. Large nuts with large kernels.
'Ridgeway'	Illinois	Very large nuts and large kernels. Heavy bearer.
'Rowher'	Uncertain origin	Mid-leafing, group B. Upright tree. Large kernels.
'Sauber'	Uncertain origin	Medium-size nuts.
'Schreiber'	Indiana	Large nuts and kernels. Heavy bearer.
'Snyder'	New York	Early season. Medium-size nuts with large kernels. Heavy cropper.
'Sol'	Indiana	Medium-size nuts, thin shells, with medium-sized kernels. Heavy, reliable bearer.
'Sparks 127'	Iowa	Late leafing, group B, early season. Medium-size nuts with large kernels. Lateral bearing.
'Sparks 129'	Iowa	Very large nuts with large kernels. Very heavy cropper.

Recommended black walnut cultivars

Cultivar	Origin	Description
'Sparks 147'	Iowa	Mid-late leafing, self-fertile. Medium-size nuts with large kernels.
'Sparrow'	Illinois	Group A, early season. Medium-size nuts with very thin shells and medium-sized kernels. Heavy annual cropper.
'Stabler'	Maryland	Mid-leafing, self-fertile. Medium-size nuts with thin shells. Peanut type.
'Stambaugh'	Illinois	Self-fertile. Medium-size nuts with large kernels.
'Stark Kwik-Krop'	Unknown	Large nuts and kernels. Heavy cropper.
'Ten Eyck'	New Jersey	Late season, productive. Nuts small, thin-shelled.
'Thomas'	Pennsylvania	Very late leafing, late flowering, self-fertile, mid-season. Medium–large nuts with good kernels. Moderate cropper.
'Thomas Myers'	Mississippi	Early season. Large nuts with thick shells and large kernels. Heavy, regular cropper.
'Throp'	Indiana	Mid-leafing, group A, late season. Peanut type.
'Todd'	Ohio	Mid-late leafing, self-fertile. Large nuts with large kernels.
'Vandersloot'	Pennsylvania	Large nuts with moderate kernels and thick shells. Heavy, regular cropper.
'Victoria'	Kentucky	Late flowering, mid–late season. Large nuts with small kernels and thin shells. Good cropper.
'Weschcke'	Wisconsin	Early season. Medium-size, thin-shelled nuts that store well.
'Wiard'	Michigan	Light cropper. Small–medium, thin-shelled nuts. Heavy cropper.

BLADDERNUTS
(*Staphylea* spp.)

European bladdernut
(*S. pinnata*)   ZONE 6, H6

American bladdernut
(*S. trifolia*)   ZONE 5, H7

Bladdernuts are a group of deciduous shrubs and small trees originating from northern temperate regions. They get their name from the fruit, which are inflated capsules containing a few nuts (seeds), and where native are found growing in deciduous woodlands in moist soil. The two species mentioned here have edible nuts and it is likely other species can be used similarly.

American bladdernut bush.

Flowers of American bladdernut in spring.

European and American bladdernuts are medium-to-large shrubs reaching 3-5m (10-16') in height and width. Their bark is smooth and striped, and they bear white flowers in spring (particularly ornamental on *S. pinnata*), borne at the shoot tips. They live for around 50 years.

European bladdernut (or 'false pistachio') originates from southern and central Europe (France to Ukraine, and Asia Minor to Syria). An upright, vigorous shrub, it bears pinnate leaves with 3 to 7 (usually 5) leaflets, each 5-10cm (2-4") long, bright green above and bluish-green below. The flowers are bell-shaped, pale whitish-green tinged pink, fragrant and about 1cm (3/8") across. They are borne in narrow drooping panicles (loose flower clusters) up to 12cm (5") long in late spring / early summer, and are pollinated by flies. Fruit are roundish-oval, pale greenish inflated bladder-like capsules up to 3-4cm (1 1/8-1 5/8") long, ripening from early autumn to early winter when they turn pale brown. They contain 2-3 (sometimes 4) glossy, roundish light brown seeds, each 8-12mm (1/4-1/2") across.

American bladdernut originates from eastern North America (Quebec to Georgia, west to Kansas and Nebraska). It is an upright, moder-

The fruit 'bladders' of American bladdernuts

ately vigorous shrub growing 3-4m (10-13') high and wide, with shiny young shoots. The pinnate leaves have 3 leaflets, each 3.5-8cm (1⅜-3¼ ") long, dark green above, downy beneath. The flowers are bell-shaped, dull white, 8mm (¼ ") across, borne in drooping panicles to 5cm (2") long in late spring / early summer; pollination is via flies. The bladder-like fruit are ridged, 3-4cm (1⅛-1⅝") long, usually 3-lobed, light green but turning brown when ripe from early autumn to early winter. They contain 2-3 glossy, roundish light brown seeds, each about 5mm (⅛") across. Established plants put up a number of suckers close to the plant.

The fruit 'bladders' of European bladdernuts

Uses

The seeds of both species are edible raw or roasted, with a pleasant pistachio flavour (hence the common name false pistachio). The shell is not edible.

A sweet edible oil can be expressed from the seeds of *S. trifolia*, used for cooking, and probably also from *S. pinnata*.

Secondary uses of bladdernuts

American bladdernut was used medicinally by the native Iroquois in North America. An infusion of plants was taken for rheumatism, and a bark infusion used as a dermatological aid. The seeds were used in child's rattles.

Plants have dense fibrous root systems and can be used for erosion control, for example on banks and steep slopes.

Cultivation

Bladdernuts are tough plants and will grow well in sun or semi-shade – one of the few nut trees tolerant of shade for forest and woodland gardens. They are fairly vigorous plants, growing some 50cm (20") per year, and are very resistant to honey fungus (*Armillaria* spp.).

Plant seedling plants at 2-5m (6'6"-16') spacing. Plants have a very dense, thick root mass and are best planted young, otherwise they can become difficult to handle.

Rootstocks and soils

Bladdernuts are usually grown from seed on their own roots. They like fertile, moist (but not waterlogged) soil and are not tolerant of drought.

Pollination

Plants are self-fertile. Flowering is more profuse following a long, hot summer previously.

Feeding and irrigation

Not required.

Pruning

Pruning is not essential. To restrict size and shape, prune after flowering. Plants can also be cut back hard in winter and will respond with vigorous growth, though you might lose crops for a year or two.

Pests

Mice and squirrels may be attracted to the ripening fruit (though I haven't had problems with either so far).

Diseases

None of note.

Harvesting and yields

Seeds ripen from early autumn to early winter over a long period. It is easy to learn to judge ripeness from the condition of the fruit 'bladder'.

Nuts of European (right) and American bladdernuts.

Fruiting starts at a young age (3-4 years) and a fully grown bush can produce 100-200 fruit in shady conditions, more in sun.

Processing and storage

These are the smallest nuts covered in this book and their size makes hand-processing tricky and potentially fiddly. Nuts of European bladdernut will go through most nut-cracking machines and this is the easiest way to release the kernels.

Propagation

Plants are usually propagated by seed. Bladdernut seed is deeply dormant and requires lengthy stratification. You can sow in autumn and pro-tect from rodents over winter – but seeds may wait until their second year to germinate. Alternatively, give seeds 13-22 weeks of warm stratification, followed by 13 weeks of cold stratification, prior to sowing in spring.

Softwood and greenwood (semi-ripe) cuttings can be taken in summer and rooted in a moist atmosphere with gentle bottom heat. Root cuttings taken in spring can also work.

Division works well after plants have suckered. Just dig out suckers in winter.

Layering can also work. Layer low branches in mid–late summer and remove rooted layers 15 months later in winter.

Cultivars

There are no named fruiting cultivars.

BUARTNUT

(Juglans x bixbyi)

ZONE 4-5, H7

Buartnuts are hybrids of two other walnut family members: the hardy butternut from North America, and the heartnut from Japan. They were first hybridized in the early 1900s in British Columbia, and later on in the century in Ontario.

Young buartnut 'Mitchell' tree.

Buartnuts combine the qualities of the two parents: cold hardiness, adaptability and sweet flavour of butternut, with vigour, more easily cracked shells and higher yields of heartnut.

In growth buartnuts resemble heartnuts, quickly becoming vigorous, wide-spreading deciduous trees up to 25m (80') high and 15m (50') wide. They live for about 100 years.

The nuts are very distinctive, most pointed at both ends, fairly rough on the outside and about 4cm (1⅝") long. Kernel percentage in improved cultivars is around 25 per cent.

Buartnut catkins.

Uses

As for butternut (page 124) and heartnut (page 166). Buartnuts have quite a high percentage of oils, approaching that of butternut, and their nutritional content is likely to be very similar to butternuts (see Appendix 1).

Cultivation

Siting / Rootstocks and soils

As for butternut (page 125).

Pollination

Buartnut is pollinated by either of its parental species. 'Mitchell' is usually self-fertile.

Feeding and irrigation / Pruning

As for heartnut (page 167).

Pests

As for butternut (page 126) and heartnut (page 167).

Diseases

Butternut canker is the most serious disease in North America – see butternut (page 127) for more details.

Harvesting and yields

As for heartnut (page 167).

Buartnut – Mitchell

Processing and storage

As for heartnut (page 168).

Propagation

Seeds need 3-4 months of cold stratification. Grafting of named cultivars is difficult and requires a hot grafting pipe or similar.

Cultivars

'Mitchell' is the most common named cultivar still propagated. See further cultivars below.

Recommended baurtnut cultivars

Cultivar	Origin	Description
'Barney'	British Columbia	Nuts large, difficult to crack; early ripening. Tree vigorous, productive.
'Butterheart'	USA	Nuts medium size, heart-shaped, crack well, kernels rich. Tree precocious (early flowering).
'Coble's No. 1'	Pennsylvania	Nuts large, quite hard to crack. Tree a slow bearer.
'Corsan'	Ontario	Nuts medium size, round. Tree vigorous, productive.
'Dunoka'	Ontario	Nuts medium and large, 25% kernels. Light crops annually.
'Hancock'	Massachusetts	Nuts medium size, of average flavour. Large, spreading tree.
'Mitchell'	Ontario	Nuts medium size, crack well, good flavour. Tree a good reliable bearer, precocious, often self-fertile.
'Van Syckle'	Michigan	Nuts large, cracks well in halves. Tree a heavy bearer.
'Wallick'	Indiana	Nuts medium size, good flavour.

BUTTERNUT

(Juglans cinerea)

ZONE 3-4, H7

Butternut is the hardiest member of the walnut family, with a range that extends well into Canada, from Georgia in the south, the Dakotas in the west, and the eastern seaboard. Trees live for about 80-90 years – less than many other walnut species. Also known as 'white walnut' and 'oilnut', due to the very high oil content, butternuts can reach 30m (100') high in their native range, but in cultivation are usually medium-size trees up to 18m (60') high. They live for about 75 years.

A young butternut tree

Straight trunks, which can reach 60-100cm (2-3') diameter, bear light grey smooth bark. All young parts of the tree, leaves and fruit are covered with fine, sticky, aromatic hairs.

The leaves are very similar to those of black walnut – compound, 35-60cm (14-24") long, with 11-19 leaflets. The leaflets themselves are serrated and pointed, 5-12cm (2-5") long and 2.5-5cm (1-2") wide, and yellowish-green. The leaves turn yellow or brown before falling in early–mid-autumn.

Male flowers are light yellowish-green catkins, mostly 5-10cm (2-4") long. Female flowers are formed in small clusters in the leaf axils, each flower with two red stigmas when ripe. Flowering is in late spring / early summer; male and female flowers on the same tree rarely mature at the same time, so self-fertility is quite limited. Cross-pollination can occur with all other walnut members but most often with heartnut, which flowers at the same time.

Each nut is borne within a green fleshy husk, which is sticky and aromatic, 4-6cm (1⅝-2½") or more long and 2.5-3.5cm (1-1⅜") across. Nuts are borne in clusters of 2 to 5. Each nut is brown and pointed, with 8 very deep, rough, sharp edges running lengthwise from top to bottom. Nuts ripen in early to mid-autumn.

Shells are hard and may require heavy-duty crackers to open. Kernels are whitish-cream with a rich oily flavour.

Uses

Butternut kernels are sweet, very oily and fragrant, with a rich, agreeable buttery-walnut flavour (see Appendix 1 for their nutritional content). Kernels comprise about 20 per cent of the total nut weight, depending on cultivar.

The nuts are eaten fresh, roasted or salted. They are used for flavouring, and are particularly popular in cooking (pastries) and confectionery manufacture. A traditional use in New England is maple-butternut candy, combining butternut with maple sugar.

Young nuts are sometimes pickled like green walnuts. They are harvested in early summer (when a pin can still be thrust through the nut without marked resistance), soaked in a mild brine for three weeks, then scalded and the outer skin rubbed off; the nuts are then covered with a 'syrup' of water, vinegar, sugar and spices.

Kernels contain about 64 per cent oils, and the expressed oil can be used as a culinary oil. Storage life is quite short.

Secondary uses of butternut

The sap can be tapped in late winter / early spring and used to make wine, or a syrup much like maple syrup.

Native Americans had many medicinal uses, mainly using the inner bark of roots. This contains juglone, juglandin and juglandic acid, and was used for treating cancer, dysentery, epithelioma, fevers, liver ailments, mycosis, tapeworms and warts. Juglone is known to have anti-fungal properties.

Native Americans and European settlers also used butternut widely as a dye source. The green nuts and bark are boiled to produce yellowish-orange (nuts) and brown (bark) dyes. These were widely used around the time of the American Civil War in 1861-1865 ('butternut jeans' became a sort of uniform for many Confederate soldiers).

The wood is highly prized by wood carvers, carpenters and interior decorators. It is satiny, warm-coloured, warpless and durable indoors, resembling black walnut but paler with chestnut-brown heartwood. Straight-grained, it has a coarse but soft texture, moderately strong and heavy, and weighs about 450kg/m^3 (28lb/ft^3). It is easily worked with both hand and power tools, and there is little resistance to cutting edges. It nails, screws and glues well and can be stained and brought to an excellent finish, but is moderately resistant to preservative treatment.

It is used for high-class joinery, interior trim for boats, superstructures, cabinet fitments, furniture, boxes and crates. It is sliced as a decorative veneer and used in place of black walnut for furniture and wood panelling. It makes a good fuel wood. Apart from North America, it has been cultivated as a timber tree in some parts of Europe.

Cultivation

Full sun is essential.

Butternuts are slower growing than heartnuts but a little faster than true walnuts. They can eventually become large, spreading trees needing 12-15m (40-50') spacing, but there is the opportunity to grow intercrops (smaller trees or other crops) in the gaps between for 10-15 years. To maximize nut crops, a fully rounded canopy is desirable, necessitating these planting distances. For good timber form, trees are planted much closer, around 5m (16') apart, but these will bear few nuts.

Although very winter-hardy, the tree is less hardy in mild temperate regions, because the cooler summers don't always properly ripen the new growth, which is then susceptible to late-spring frost damage.

As in many of the walnut family, the substance juglone occurs in roots and is washed into the soil from decaying leaves. Juglone is quickly detoxified by the soil, but in some circumstances and soils it may rise to concentrations that have growth-supressing effects. These can be severely detrimental to apples in particular, but also affect the heather family (Ericaceae), cinquefoils (*Potentilla* spp.), eastern white pine (*Pinus strobus*) and red pine (*P. resinosa*), potatoes and tomatoes, and French beans. These species should be avoided in planting schemes close to butternuts.

Rootstocks and soils

Butternut is usually grown from seed, or grafted on to black walnut, butternut or walnut rootstocks.

Like the rest of the walnut family, butternuts prefer a deep, fertile soil of near-neutral pH (ideally pH 6.0 to 7.0), moist but also well drained, although they tolerate a high water table. Limestone soils are tolerated.

Trees resent transplanting, as they produce a deep taproot. They are best transplanted as young plants – older plants may take a year or two to recover and grow little during that time.

Male butternut flowers.

Female butternut flowers.

Pollination

As with other walnut family trees, butternuts are wind-pollinated and most trees are not very self-fertile. A different butternut or a heartnut will usually cross-pollinate satisfactorily.

Feeding and irrigation

Little feeding is needed. Butternuts haven't been bred and selected for the heavy cropping seen in true walnuts, and the crops don't require high fertility. Growth is about 3m (10') in 10 years in British conditions; rather more in a warmer climate.

Butternuts are deep-rooted trees and rarely require irrigation.

Pruning

Little pruning is required. Formation pruning in the first few years to remove branches with small (acute) branch angles is about all that's needed. Cropping trees are not usually pruned.

Pests

North American minor insect pests include the walnut caterpillar (*Datana integerrima*) and the fall webworm caterpillar (*Hyphantria cunea*), which attack leaves; and the butternut curculio or walnut weevil (*Conotrachelus juglandis*), whose larvae feed on young stems, branches and immature fruit. The latter can be serious in Canada, where a feeding deterrent is sometimes sprayed on trees. Walnut husk flies (a problem on walnuts in North America) don't usually attack butternut.

Minor pests can include the red spider mite (*Panonychus ulmi*) and the European fruit lecanium (*Lecanium corni*, syn. *Parthenolecanium corni*).

Squirrels and crows are sometimes a problem. In North America, the purple grackle can be a pest, often destroying immature fruit by pecking at the green husks.

Diseases

In North America butternut canker is a serious disease, caused by the fungus *Sirococcus clavigignenti-juglandacearum*. Symptoms are dying branches, discoloured bark, and cankers on twigs, branches and trunk. Young cankers appear sunken, dark and elongated, and ooze a thin black liquid in spring. Older cankers are large and may be covered with shredded bark. Several cankers may coalesce and girdle a tree, causing its death. This disease is decimating much of the butternut population in its native range, but resistance is occasionally occurring.

Walnut leaf blotch (*Gnomonia leptostyla*) can be a problem in humid climates, but is less severe than on true walnut. See walnuts (page 272) for more details. Walnut blight does not attack butternut.

Walnut bunch is caused by a mycoplasma-like (virus-like) organism, resulting in witches' brooms (clusters of wiry twigs on branches). Butternuts are quite susceptible to this in North America.

The fungus *Melanconis juglandis* can cause a slow dieback of branches, with no well-defined symptoms. Trees growing weakly are more susceptible.

Harvesting and yields

Nuts ripen early to mid-autumn – earlier than true walnut. They will fall when ripe and/or can be knocked off, usually within the green husk, although this may partly fall off.

Trees can take 6-10 years before they start to crop, and yields can reach a maximum average of about 25kg (55lb) per mature tree (in-shell).

Processing and storage

Any husks must be removed before they blacken, as this will affect the nuts. See black walnuts (page 110) for tips on husk removal. Butternut husks are gummy and result in gummy hands and gloves. Leather gloves should be worn to protect fingers from the sharp ridges of the shells.

Drying is necessary for storage more than a few weeks. Blown-air-drying is required in a cool, humid climate.

Butternuts contain very high oil levels – more than most nuts – which makes them harder to store. They will store for a few months in room temperatures and up to a year or so at fridge temperatures if properly dried.

Use a heavy-duty nutcracker to crack the nuts. To make cracking easier, they can be covered with hot water and soaked for a couple of hours prior to cracking.

Propagation

Seeds require 3-4 months of cold stratification. Germination may be improved by carefully cracking the shells before sowing. Autumn sowing can be effective, but make sure that rodents can't get at the seeds. Expect up to 50 per cent germination rate. On average there are 66 seeds/kg (30 seeds/lb). Seedlings of named varieties will inherit many of their qualities.

Cultivars are propagated by grafting on to black walnut or true walnut rootstock. Grafting is difficult and may require the use of a hot grafting pipe or alternative system for keeping grafts warm; techniques used include splice grafts, chip budding and greenwood (semi-ripe) tip-grafts. You'll find detailed information on these techniques in R. J. Garner's *The Grafter's Handbook* (see Resources).

Cultivars

Most cultivars are suited to cool climates. Little breeding work has been done on butternut, so seedling trees have a good chance of being as good as the parent trees. Grafted trees are available from a very few nurseries in North America and Europe, and will give predictable results – especially important if you only want to plant a few trees.

Improved cultivars have been selected for thinner shells (thus easier and improved cracking) and larger percentage of kernel. Most selection took place in the first half of the 20th century by members of the Northern Nut Growers Association (NNGA – see Resources). Some selections have been lost but those still cultivated include 'Bear Creek', 'Beckwith', 'Craxezy' and 'Kenworthy' (see table opposite).

Butternut – Bear Creek

Butternut Beckwith

Butternut – Booth

Recommended butternut cultivars

Cultivar	Origin	Description
'Ayers'	Michigan	Nuts medium size, crack well, high percentage of kernel. Tree vigorous, upright, late leafing and flowering, resistant to walnut leaf blotch.
'Bear Creek'	Washington	Nuts medium size, crack very well. Tree a heavy bearer.
'Beckwith'	Ohio	Nuts medium size, crack quite well. Tree moderately vigorous, a prolific annual cropper, resistant to walnut leaf blotch.
'Booth'	Ohio	Nuts medium size, crack well. Tree vigorous, moderately susceptible to walnut leaf blotch.
'Bountiful'	Missouri	(Stark's Bountiful) Nuts mild flavoured, easily cracked and shelled. Tree a heavy cropper, self-fertile, flowers frost-resistant.
'Buckley'	Iowa	Nuts very large, crack quite well, and kernels are of good quality. Tree very vigorous, early leafing, has some resistance to walnut leaf blotch.
'Chamberlin'	New York	Nuts medium–large, crack moderately well; kernels moderately well filled, good quality. Tree moderately vigorous, susceptible to walnut leaf blotch and dieback.
'Craxezy'	Michigan	Nuts medium size, easily cracked, well filled, and kernels are of good quality. Tree yields well, early leafing, moderately susceptible to dieback, has some resistance to walnut leaf blotch.
'Creighton'	Michigan	Nuts small / medium size, crack very well and well filled; late ripening. Tree vigorous, late to leaf out and lose leaves in autumn, resistant to walnut leaf blotch.
'Fort Wood'	Missouri	Nuts hard to crack. Tree productive, easily grafted.
'George Elmer'		Nuts medium size, rounded, good quality, crack well. Tree vigorous, susceptible to walnut leaf blotch.
'Gray Road'	Indiana	Nuts large, crack moderately well.
'Henderson '#1' & '#2'	Illinois	Selections with very good timber form.
'Iroquois CA'	Ottawa	Useful timber selection. Selected for butternut canker resistance – not nut qualities. Medium size nuts.
'Kenworthy'	Wisconsin	Nuts large, crack well, good flavour. Tree small, a heavy bearer, precocious (early flowering), resistant to walnut leaf blotch.
'Kinneyglen'	New York	Nuts medium size, crack very well, well filled.
'Love'	Michigan	Nuts small, crack well, good quality. Tree vigorous, precocious, resistant to walnut leaf blotch.
'Mandeville'	New York	Nuts superior, crack well.
'Moorhead #1'	Kentucky	Useful timber selection. Selected for butternut canker resistance – not nut qualities.
'My Joy'	Pennsylvania	Nuts medium size, crack very well, well filled.
'Painter'	Pennsylvania	Nuts very large.
'Sherwood'	Iowa	Nuts are well filled and kernels are of good quality. Precocious tree.
'Weschcke'	Wisconsin	Nuts medium / large size, crack well, well filled; kernels are light-coloured and of good quality.

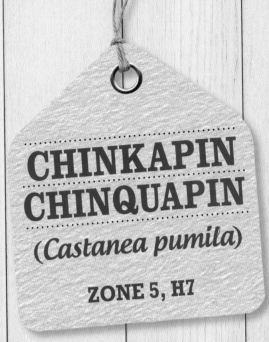

CHINKAPIN
CHINQUAPIN

(*Castanea pumila*)

ZONE 5, H7

Also known as dwarf or bush chestnuts, chinkapins are deciduous shrubs and small trees found throughout eastern, southern and south-eastern USA.

Botanists have had fun with chinkapins: some have separated them into eight or more poorly defined species, but there is now general agreement that all these species come within *C. pumila*. Here I have concentrated on the most common form, var. *pumila*, known as the Allegheny, American, common or tree chinkapin.

Chinkapin leafing out in spring in the author's forest garden.

Chinkapins are multistemmed suckering shrubs or trees growing 2-4m (6'6"-13') high, with spreading habits and smooth bark. In their native range, occasional single-stemmed plants may reach twice as high (to 8m/26') and are found on dry sandy soils in woods and thickets. Their lifespan can be 100 years or more.

Leaves are borne alternately along the slender reddish-brown shoots. The leaves are typical chestnut shape but very variable in size, 7-15cm (3-6") long and 2-5cm (¾-2") wide; they are bright, shiny yellowish-green to light green above, whitish and downy beneath, and have serrated margins.

Flowers are borne from leaf axils of the current year's shoots. Male catkins appear near the bases of the shoots, and are 10-15cm (4-6") long. Female flowers are small, like miniature burrs. Chinkapin often produces bisexual flowers containing both male and female flowers on stalks, with females near the base and males near the tips. The flowers are strong-smelling.

Each nut develops within a softly prickly burr, 14-46mm (⅝-1⅞") in diameter. Each flower spike develops into 1 to 5 (sometimes up to 8) burrs. The burrs split when the nuts are ripe and a few days later the nuts fall out. The nuts are almost spherical, shiny brown, some 12-20mm (½-¾") in diameter. There are 500-1,300 nuts/kg (225-590 nuts/lb).

Uses

Chinkapin nuts are sweet, aromatic and nutty flavoured. They can be used in all the same ways as sweet chestnut, though they are more palatable raw, as they contain more sugars (sucrose

Chinkapin nuts are a beautiful shiny brown.

and glucose) than sweet chestnut. They contain higher levels of oleic and linoleic fatty acids than European chestnuts (see Appendix 1 for their nutritional content).

After drying, the nuts can be ground into flour (gluten-free) and used to make pancakes, added to bread mixes, etc. Roasted nuts can be used to make a coffee substitute.

Secondary uses of chinkapin

Pannage systems are sometimes used with pigs fattened on chinkapins, particularly in wild stands. Dense thickets of chinkapins make good cover for birds such as quail.

The wood is a good source of fuel and makes good charcoal. Bushes coppice well. The wood is naturally durable, strong, dark brown, but is rarely available in quantity.

Like sweet chestnut, the leaves, bark, wood and seed burrs all contain tannins (yellowish or brownish bitter-tasting organic substances) and could potentially be used for tanning leather.

It has potential as a dwarfing rootstock for other *Castanea* species, though the suckering habit could be a problem, and there have been incompatibility issues.

Chinkapin has a number of traditional medicinal uses. The leaves have been used as a dermatological aid and febrifuge (reducing fevers); also as an antiperiodic (preventing recurring attacks of a disease), astringent and tonic. The root has been used as an astringent, a tonic and to treat fevers.

Cultivation

Chinkapin requires full sun to crop well, though it will tolerate light shade. Shelter is also required to stop the pollen being blown away too quickly in summer.

Plant at 2-4m (6'6"-13') apart, the larger spacing allowing for a full, rounded canopy to develop to maximize the crop.

Rootstocks and soils

Like other members of the sweet chestnut family, chinkapin prefers a well-drained soil that is somewhat acid (pH 5.5 to 6.0 is ideal). Trees are normally grown from seed on their own roots.

Pollination

Flowering takes place in midsummer and begins when bushes are 2-4 years old. Flowers

Chinkapin flowers. Note the bisexual flower (top left) with female flower at the stalk base.

are smaller versions of those on sweet chestnut: male flowers are catkins, female flowers are tiny and burr-like.

Flowers are mainly wind-pollinated, but insects (mainly bees) may also contribute to pollination. Most chinkapins are not self-fertile, so two seedling bushes should be grown near each other for pollination.

Feeding and irrigation

Chinkapins are not heavy feeders, but if crops are large, then some added nitrogen can be beneficial.

Plants are quite drought-tolerant and irrigation is unlikely to be necessary in most climates.

Pruning

Little pruning is required. Chinkapins are fairly finely branched and the canopy can be allowed to develop naturally. Some pruning of lower branches for the first few years as the bushes grow is desirable, to allow for access underneath for harvest and possible underplanting/management.

Pests

The main pests are squirrels. Crows can also be a problem on occasion.

Diseases

Like sweet chestnuts, chinkapins can suffer from ink disease (*Phytophthora cinnamomi* – see page 234), especially on soils which are not well drained. Try to improve drainage if necessary before planting.

They have some resistance against chestnut blight (*Cryphonectria parasitica* – see page 234) and have been used in some American chestnut breeding programmes as a source of resistance. If they do get chestnut blight, their naturally suckering habit means that they can in effect tolerate the disease and still crop well.

Chinkapin is resistant to honey fungus (*Armillaria* spp.).

Harvesting and yields

The nuts will start to fall when ripe in mid- to late autumn. The ripening period for a single bush is over several weeks, with basal burrs ripening earliest. The old burrs usually remain on the bushes for weeks or months, and so interfere little with harvest.

It is well worth shaking the branches to bring down loose nuts – the branches are thin and easily manipulated. Place a tarp or sheet on the ground to collect the nuts, which are just too small to be collected efficiently with a hazel-size Nut Wizard.

In very wet autumns, nuts can start to germinate within the burrs, similar to sweet chestnut.

Yields can reach about 6kg (13lb) per bush or up to 3t/ha (6,500lb or 2.9t/acre).

Processing and storage

Dry the nuts using blown air at about 40°C (104°F). They take 1-2 days to dry. Once properly dry they will store for years, like sweet chestnuts.

Chinkapin.

Propagation

Nuts should be sown in autumn immediately after harvest. Keep moist but not wet over winter. They will germinate immediately if the soil or compost is still warm, putting down a taproot first, with top growth only starting in late winter or spring.

Suckers can easily be removed from established thickets.

Cultivars

There are no named cultivars.

GINKGO MAIDENHAIR TREE

(Ginkgo biloba)

ZONE 4 H7

Originating from China, though now only found in the wild in the Tianmu Mountain area of Zhejiang province, the ginkgo has been widely grown for a very long time in China, Japan and Korea. It is a long-lived tree, which is one reason why it is primarily found around temples in Japan, Korea and Manchuria, where it is regarded as a sacred tree.

The distinctive fan-shaped leaves of ginkgo.

The name 'ginkgo' derives from the Chinese *yin-kuo*, via the Japanese pronunciation *gink_*. The tree was introduced to Japan over 1,000 years ago, and the Japanese name for the species translates as 'silver apricot'.

Also called maidenhair tree, ginkgo is a relic of prehistoric ages, being the only survivor of a genus that was widely distributed 180 million years ago. The tree was introduced to Europe in around 1730, from seeds collected from trees in temple gardens.

Ginkgo is a tall deciduous/coniferous tree, growing up to 25m (80') in cool climates, though up to 40m (130') in warmer climates. Trees are usually narrow and upright for several decades, but can become spreading with age. Trunks often branch low, with several erect main branches forming. The bark is grey and deeply furrowed on older trunks. Trees can live to a great age, sometimes over 1,000 years. They often live for 200 years or more in cultivation.

The brown buds break in late spring. Shoots are of two types: long extension growths with alternate leaves; and short woody spurs bearing fruit and leaves in false whorls of three to five. Ginkgo leaves are distinctive and unmistakable: fan-shaped, with branching parallel veins and an irregular undulating margin, leathery and tough, 5-12cm (2-5") wide. They are paler beneath and usually cut into two lobes. The leaf stalks are relatively long at 2-9cm (¾-3½").

Vegetative growth ceases in late summer.

Male flowers of ginkgo.

136

Female flowers of ginkgo.

Leaves turn a beautiful golden-yellow in autumn before quickly falling.

Flowers appear in late winter / early spring before the leaves fully open from the leaf axils. Male flowers are yellow, catkin-like, pendulous, solitary, and 25-80mm (1-3¼") long; female flowers are pale yellow, becoming orange, tiny (2mm/¹/₁₆"), and appear in twos or threes on long stalks. Pollination is via the wind and female flowers develop into drupe-like fruit, with a fleshy outer part and hard inner nut.

The ginkgo is dioecious, so trees are either male or female; only females bear fruit. There is no sure way of telling male and female seedlings apart before they flower, though there are three unsubstantiated ways of telling the sexes apart:

1) It is believed that males leaf out 2 weeks earlier in spring and drop their leaves 2 weeks later than the females in autumn.
2) Seeds are marked by 2 or 3 longitudinal ridges. Traditional Chinese sources say that those with two produce females, while those with 3 produce males.
3) One source says that females are almost horizontally branched, with deeply lobed leaves, while males branch at a sharper angle with less lobed leaves.

Flowering may not occur until trees are 20-30 years old, hence there is a long wait for nuts!

The fruit appear singly or in pairs, like small round yellowish-green plums, 25mm (1") long. The inedible fleshy exterior starts to decay on

137

Ginkgo fruit.

Nuts are first shelled – the shells are thin and easily removed, especially if nuts are boiled in water for a few minutes first. The kernels are then usually soaked in hot water to facilitate peeling off the papery pellicle, which (as in most nuts) can be bitter.

Ginkgo seeds are safe to eat cooked in moderation (they contain ginkgotoxin, which raw / in quantity can cause a number of unpleasant side effects). Kernels are eaten roasted or boiled for 10-15 minutes, when they have a pleasant and characteristic flavour, which has been likened to almond or mild Swiss cheese. Eaten at Chinese feasts, they are supposed to aid digestion and alleviate the effects of drinking too much wine. Tinned, boiled ginkgo nuts can sometimes be seen in oriental grocery stores. The Chinese often bake them with meat or fowl and include them in sweet soups with Chinese dates (*Ziziphus jujuba*) or white fungi.

the tree, darkening to purple-black; ripe fruit fall in autumn, the fleshy covering then bursting and emitting an unpleasant odour of rancid butter. The inner nut is smooth and white, ovoid, 12-20mm (½-¾") long, with 2 or 3 ridges.

Uses

When cooked the nut kernel is well flavoured and esteemed as a delicacy in China and Japan, where ginkgo nuts are sold in markets and mostly used on special occasions such as weddings and the Chinese New Year.

Dry kernels constitute about 59 per cent of the whole nut weight. Ginkgo is a starchy nut, as Appendix 1 shows, which also contains oleic, linoleic and palmatic fatty acids.

Ginkgo biloba

Ginkgo nuts.

138

Secondary uses of ginkgo

The oil from the kernels is also edible when cooked, and it can also be burned as an illumination source. However, the yield is low – just over 3 per cent.

Ginkgo has traditionally long been used as a street tree in Japan, and more recently in other countries. Male cultivars are usually used, so there is no unpleasant-smelling fruit for folk to complain about – the fruit flesh also becomes slimy and slippery as it decays, so not ideal on pavements.

The timber is not particularly highly prized – it is yellow-brown, light, soft, close-grained and brittle, with a thin satiny-white sapwood. It has insect-repelling properties and is used for chessboards, bas-relief carvings and toys. Trees coppice well, but there would be little point in doing so unless cultivating for the medical leaves (see below.)

Extracts from the leaves and roots (with alcohol) have been found to have effective pest-control properties. Leaf extracts act as a repellent to silverfish and as an anti-fungal to brown rot of pome fruit (*Monilinia fructicola*); root extracts act as an antifeedant and insecticide to the European corn borer (*Ostrinia nubilalis*), and as an antiviral to the southern mosaic bean virus.

Traditional Chinese medicinal uses of ginkgo

The fruit pulp, seeds and leaf extracts have long been used in traditional Chinese medicine. The fruit pulp is macerated in vegetable oil for 100 days and used for tuberculosis and pulmonary complaints. Raw nut kernels are antitussive (prevent/relieve coughs), astringent and sedative, and are said to have anti-cancer properties; they are used for bladder ailments, cardiovascular ailments and cancer. The kernels are also sometimes used fried. Leaf extracts are used in peripheral arterial circulation problems, and are inhaled for ear, nose and throat ailments.

Modern medicinal uses

The numerous uses of ginkgo in Chinese medicine brought the plant to the attention of mainstream medicine. Leaf extracts have an antiradical effect on the brain – i.e. aid the body to repel attacks of free radicals, which damage cell membranes – and are vasoregulatory (improving general microcirculation, especially in the brain). It is also thought to slow down cerebral ageing by improving the glucose consumption of the brain. The effects are to improve performance of short-term memory, alertness and drive; other uses are for cerebral oedema, tinnitus, dizziness, dementia and Parkinson's disease, etc. Leaf extracts are now used in a drug to treat senility, while another extract is used for sufferers of diabetes mellitus and arterial disease. Some large pharmaceutical companies have invested in ginkgo plantations to provide bulk leaf matter for drug extraction.

Ginkgo.

Cultivation

Ginkgos are tolerant of most conditions, including part shade, steep slopes, high and low humidity, air pollution (hence use as a street tree), and are very drought-tolerant. The two things they dislike are poor drainage and great exposure.

Plant at 4-10m (13-33') spacing. Female trees should not be too far (within 50m/160') from a male pollinator to ensure pollen transfer. If possible, place a male on the main windward side of females so that pollen gets blown in the right direction. If you have room for only a single tree, then it is feasible to graft a male branch on to a female tree for pollination purposes.

Growth is moderately fast in cool climates – about 3m (10') in 10 years and 8m (26') in 20 years – although faster in climates with hotter summers.

Rootstocks and soils

Named cultivars are grafted on to seedling rootstocks.

Any soil is acceptable, as long as it is fairly well drained. Preferred conditions are a deep, well-drained soil with pH on the acid side (5-5.5). Ginkgos tolerate acid soils (to pH 4.5) and alkaline soils (to pH 8.5).

Pollination

Ginkgo flowers are wind-pollinated. Flowering is early in the season, so poor wet or cold weather can have a serious effect on crop set.

One male tree should pollinate several females in the nearby vicinity.

Feeding and irrigation

Rarely required. Ginkgos are fairly drought-tolerant and the normal yields don't require regular feeding.

Pruning

Little, if any, pruning is required.

Pests/Diseases

Ginkgos are hardly ever affected by pests or diseases, and are immune to honey fungus (*Armillaria* spp.).

The leaves turn a beautiful rich yellow in autumn.

Harvesting and yields

Trees start bearing fruit at 25-35 years of age. The ripe fruit fall or can be shaken from trees, or beaten down with a bamboo pole, as in China. The fruit flesh must be removed to harvest the nuts – this is the most challenging part of ginkgo cultivation, as the fruit flesh, once crushed or damaged, smells unpleasant/offensive. Early-picked fruit will not have properly developed kernels.

Processing and storage

The decaying fleshy fruit exterior has been variously described as 'unpleasant', 'evil', 'foul', 'offensive' and 'malodorous'. This is caused by butanoic acid, which causes the smell of rancid butter. The fruit are kept in a container or piled up outdoors, to allow the pulp to ferment, after which nuts are easily removed and washed clean.

Gloves should be used when handling fruit, as the juice from the outer flesh causes itching, rashes and even dermatitis in some people. Also ensure good ventilation when handling the fruit, as some people complain of headaches from long periods of smelling it!

Once nuts are separated, they can be surface-dried and will then store for a year or more in fridge conditions (5°C/41°F).

Propagation

Ginkgo grows easily from seed. No stratification is required and seeds can be sown in autumn or spring. There are about 600 seeds/kg (270/lb). Sow into deep cell containers in outside seedbeds at a density of 300 seeds/m² (250/sq yd) and take precautions against rodents and birds. The germination rate is usually quite high but can be spread over several months. Late-germinating seedlings are susceptible to frost damage in early autumn, but there are generally no pest or disease problems. First-year growth is 20-40cm (8-16").

Named cultivars are usually grafted on to seedling rootstocks using any normal method in spring.

Cultivars

Good fruiting cultivars have been selected in China and Japan, and a few of these are available in Australia, Europe and North America (see table below). Several ornamental cultivars are also available, mainly selected because of their leaf shape or tree form (mostly columnar or narrowly conical). Some of these are of known sex, so females can still be used for nut production.

Self-fertile selections do occur on occasion in the wild. One was named in the 1990s: 'Dr Causton' in the UK.

In China, cultivars are divided into three groups:

1) Meihe-Yinxing group or plum-stone-shaped ginkgo, with round fruit;
2) Fushon-Yinxing group or finger citron ginkgo, with elliptic or oblong fruit; and
3) Maling-Yinxing or horse's-bell-shaped ginkgo, with a fruit shape intermediate between the other two, and with a small point on the top of the fruit.

Recommended ginkgo cultivars

Male cultivar	Origin	Description
'Autumn Gold'	USA	Pyramidal tree when young, spreading with age.
'Fairmount' (syn. 'Fairmont')	USA	Tall columnar tree with a dense crown and narrow leaves.
'Lakeview'	USA	Pyramidal tree.
'Mayfield'	USA	Very narrow and upright tree.
'Palo Alto'	USA	Pyramidal tree.
'Princeton Sentry'	USA	Tall, columnar tree.
'Saratoga'	USA	Small tree up to 4m (13') high with a dense compact habit and narrow leaves.

Female cultivar	Origin	Description
'Eastern Star'	China	Heavy cropping with large nuts.
'King of Dongting Mountain'	China	Large leaves, medium-size tree.
'Long March'	China	Heavy cropping with large nuts.
'Ohazuki' (syn. 'Epiphylla', 'Ohatzuki')	Japan	Tree only 4-5m (13-16') high with large leaves.
'Variegata'	unknown	Slow-growing tree up to 3m (10') high with variegated leaves. Prefers part shade.

GOLDEN CHINKAPINS

(*Castanopsis* spp.)
(*Chrysolepis* spp.)

ZONE 6-9, H4-6

Golden chinkapins (also called golden chestnuts) are evergreen trees and shrubs from milder regions: *Castanopsis* from mainly subtropical southern and eastern Asia, *Chrysolepis* from western USA. They are intermediate in character between oaks and sweet chestnuts, and all have acorn-like nuts that can be used like oak acorns. Most have a lifespan of less than 100 years.

Golden chinkapin (*Chrysolepis chrysophylla*).

Ripening nuts on Japanese chinkapin (*Castanopsis cuspidata* var. *sieboldii*).

Flowers on Dwarf golden chinkapin (*Chrysolepis sempervirens*).

Uses

Golden chinkapin nuts are eaten, usually cooked and often after removal of tannins (as with oak acorns – see page 192). Their nutritional content is similar to oak acorns (see Appendix 1).

Secondary uses of golden chinkapins

The timber from larger shrubs or trees is valued for fuel and fencing as well as construction uses. Logs from some Asiatic species are used to cultivate shiitake mushrooms.

Leaves are sometimes used for fodder, which may be particularly useful in winter. Branches can be lopped and given to animals to strip.

Cultivation

Castanopsis prefer a warm continental climate and only flourish with a good amount of summer heat. Although winter-hardy in some cool temperate climates, the summers are not good enough for them to thrive.

Chrysolepis like both Mediterranean and oceanic climates. They do very well in cool temperate climates.

Planting distances vary, depending on size of the species (see the table on page 146). Leave enough space for the tree to grow full size.

Rootstocks and soils

Golden chinkapins are usually planted as seedlings. Like sweet chestnuts, they like an acid to neutral soil that is well drained. Soils that get waterlogged should be avoided, as in these soils plants may die from root asphyxiation in winter.

Pollination

These species are adapted for wind pollination, like sweet chestnut. However, bees and other insects are attracted to the flowers and can aid pollination and nut set. It may be worth encouraging bees into the orchard with other flowering plants.

Feeding and irrigation

Neither is usually required.

Pruning

Little, if any, required.

Flowers are formed on younger growth so can decline on old trees or shrubs. In this case, cutting back may stimulate new growth.

Pests

There are no significant pests.

Diseases

Both families are resistant to chestnut blight (*Cryphonectria parasitica*).

Harvesting and yields

Harvest the nuts in the second autumn after flowering. The spiny burrs will split and the nuts fall to the ground, so make sure any grass or plant growth underneath nut plants is cut short. You can also place nets or tarps below trees to catch the nuts.

Nuts of Japanese chinkapin (*Castanopsis cuspidata* var. *sieboldii*).

Processing and storage

Treat like oak acorns (see page 190).

Propagation

Normally propagated by seed, which isn't dormant and should be sown immediately in autumn. Grafting on to oaks or sweet chestnuts may be possible.

Species

There are no named cultivars but the species listed on page 146 all have potential. Note that in cool temperate climates, trees do not grow to the sizes reached in their native climes, and plants are often more shrub-like. Species listed with their hardiness in brackets are little known and their hardiness is not verified.

Recommended golden chinkapins species

Species	Origin	Hardiness	Description
Castanopsis chinensis	China (Yunnan, Kwantung up to 1,300m/ 4,265' high)	Z7/H5	A tree up to 15-18m (50-60') high or more. Leaves 8 x 3cm (3¼ x 1⅛"), glossy green above and bluish beneath. Bears densely prickly greenish-brown fruit in clusters 10cm (4") long.
Castanopsis cuspidata	Japan	Z7/H5	Japanese chinkapin. A tree up to 15m (50') high or more with wide, drooping branches. Leaves, 5-9 x 2-4cm (2-3½ x ¾-1⅝"), are dark glossy green above, pale metallic brown and scaly beneath. Fruit clusters are 5-7cm (2-3") long with 6-10 fruit, round at first but oval when ripe, 1.5cm long by 1cm thick (⅝ x ⅜"), hairy but not prickly, each enclosing one nut. Var. *sieboldii* (syn. *C. sieboldii*) has larger nuts but may be less hardy.
Castanopsis delavayi	China (Szechwan, Yunnan up to 3,000m/9,843' high)	Z8/H4	Tree up to 15m (50') high or more. Leaves 7-14cm (3-6") long, silvery-grey or whitish beneath. Male catkins are yellow, slender, and 10-18cm (4-7") long. Fruit with short prickles are borne in clusters of 6-10, each round, 12mm (½") thick, with one nut.
Castanopsis hystrix	Eastern Himalayas	(Z8/H4)	Tree up to 20m (70') high, which is found as high as 2,500m (8,200') in the Himalayas. Tree is also lopped for fodder.
Castanopsis indica	Himalayas	(Z8/H4)	Indian chestnut. Found at up to 2,500m (8,200') in the Himalayas. Tree up to 15m (50') high.
Castanopsis orthocantha	China (Yunnan, Szechwan up to 2,600m/8,530' high)	Z8/H4	A tree up to 10-20m (33-70') high. Bluish-green leaves, 6-12 x 2.5-4cm (2½-5 x 1-1⅝"). Has very fragrant male flowers, 8-10cm (3-4") long. Fruit clusters are 6cm (2½") long, with burrs rounded, 3cm (1⅛") across.
Castanopsis sclerophylla	China (Hupeh up to 1,500m/4,921' high)	(Z8/H4)	Tree with reddish-brown shoots. Fruit clusters of 1-3, each bearing a nut 10-14mm (⅜-⅝") across. Nuts sweet, eaten raw or cooked. Leaves are used medicinally.
Castanopsis tibetana	Tibet	(Z8/H4)	Tibetan chinkapin. Bears nuts 2cm (¾") or more across.
Castanopsis tribuloides	Western China, Himalayas (up to 2,200m/7,218' high)	(Z8/H4)	A tree usually 5-10m (16-33') high, occasionally more. Bears nuts about 1.5cm (⅝") across. The plant is lopped for fodder.
Chrysolepis chrysophylla	Western USA	Z7/H5	Golden chinkapin, golden chestnut. A tree up to 25m (80') high with furrowed bark. Leaves 4-15 x 1-4cm (1⅝-6 x ⅜-1⅝"); the spiny husk contains 1-2 seeds, 1cm (⅜") long. Seeds are sweet and eaten raw or cooked. The wood is light brown, fine-grained, light, soft, not strong – used for agricultural implements, furniture, panelling, veneer, cabinet work and fuel. Var. *minor* is a smaller shrub or tree up to 10m (33') high, originating from the coast of California, and is less hardy – to Z8/H4.
Chrysolepis sempervirens	Western USA	Z6 / H6	Dwarf golden chinkapin. Tree or shrub reaching 3-5m (10-16') high and 6m (20') wide with smooth bark. Leaves 2-6cm (¾-2½") long. Also known as bush or Sierra chinkapin, the nuts are sweet and eaten raw or cooked.

HAZELNUT and FILBERT

(*Corylus avellana, C. maxima*)

ZONE 4, H7

Hazelnuts and filberts are deciduous large, woody shrubs or small trees originating from Europe. Although in theory filberts are of the species *C. maxima* and hazelnuts (or cobnuts) of the species *C. avellana*, the names have become confused over the years and some hazels are called filberts and vice versa. Since the two species are very similar in all respects, it makes no sense in this context to separate them, so from here on I will just refer to hazels or hazelnuts, but everything applies to filberts as well.

A young hazel tree being trained with a single stem.

Hazel is native to western Asia, northern Africa and most of Europe, from the western margins to Scandinavia and the Mediterranean. It is usually found in hedges and in lowland or upland deciduous woods as part of the understorey, and is frequently with alder, ash, birch and oak. Filbert is native to south-eastern Europe (the Balkans) and Asia Minor, but is widely naturalized elsewhere.

C. avellana was one of the first species to recolonize Europe (including the British Isles) after the last ice age. Pollen counts from peat bogs covering the period 8000 to 5500 BC show that hazel pollen exceeded that of all other species together. Nuts from these species have been widely used since prehistoric times; remnants have been found in Mesolithic sites, and it's been suggested that they provided a staple source of food, which was later to be taken by cereals. Hazelnuts were found in the ruins of Pompeii, and Charlemagne is known to have spread the species throughout Europe – six cultivars were already being grown in Italy in 1671. Hazelnuts are now a major commercial worldwide crop.

The term 'filbert' probably originates from the French Saint Philbert, whose feast day was celebrated on 20 August. The word 'hazel' originates from the Anglo-Saxon word for 'hood' or 'bonnet' – *haesel*. Other common names for hazel include cobnut, Pontic nut, Lombardy nut and Spanish nut.

Hazelnuts and filberts are mostly large, multistemmed shrubs, growing 5-6m (16-20') high and wide, though sometimes more tree-like and higher. The stems have mottled grey and brown bark, and show prominent breathing pores.

The leaves open in mid-spring, are 6-10cm (2½-4") long (a little wider in filbert) and have 6-8 vein pairs, stiffly hairy and almost circular, with double-toothed edges and a short pointed tip. They fall late, in early winter.

Most people know the male flowers well: yellow catkins 3-7cm (1⅛-2¾") long, ripe and shedding their pollen for about 6 weeks, varying between mid- to late winter on different trees. The female flowers are much less noticeable, being tiny, 5mm (⅛") with red tassels. Trees are not self-fertile. Pollination is via the wind, so weather conditions in late winter, when flowering occurs, have a big impact on nut set.

Clusters of 1 to 4 nuts form in summer and autumn, each nut surrounded by leafy bracts. In hazel, the leafy bracts are shorter than the length of the nut – i.e. the nut sticks out from the bracts. In filbert, the bracts are tubular, surrounding the whole nut, and virtually closed at the end, enclosing the nut.

The nuts are pale green ripening to pale brown. In hazels, they are round, oval, egg-shaped or even conical. Filberts tend to be longer and narrower, sometimes almost cylindrical. Nut length can vary from 15 to 30mm (⅝-1⅛"). Nuts ripen over a period of 4-6 weeks, early to mid-autumn in cool regions but from late summer in warmer climates.

Hazels have small taproots and plentiful shallow roots. Their normal lifespan is 70-80 years at most. Hazels often form mycorrhizal associations with fungi (see Chapter 1, page 27).

Uses

Hazelnuts are eaten raw, roasted or salted, and are used in many confections and baked goods – see Appendix 1 for their full nutritional content details. Hazels are an important commercial crop in many parts of the world.

A clear yellow non-drying oil is pressed from the nuts. Having a strong, delicious hazel flavour, this is used in salad dressings and for cooking food; it can also be used for painting, in perfumes, as a fuel oil, for the manufacture of soaps and for machine lubrication.

Secondary uses of hazelnut and filbert

Hazel is well known as a coppice species (though coppice is not ideal for nut production – see 'Pruning', page 151). The poles from coppice ('wands') are long and flexible and have been traditionally used for many years for wattle fencing (branches are usually split, then weaved to make sections of fence); water diviner's rods; thatching spars; walking sticks; fishing rods; basketry; clothes props; pea and bean sticks; hedging stakes used when laying a hedge; firewood, notably for brick kilns and baking ovens; construction of wattle-and-daub walls; crates; hurdles; barrel hoops; and fascines for laying under roads in boggy areas. Hazel wood is soft, elastic, reddish-white with dark lines, is easy to split but not very durable. Joiners and sieve makers sometimes used older wood, and charcoal from the wood was a component in gunpowder manufacture. Root wood, veined and variegated, was once used for inlay cabinet work.

Hazels are an excellent hedging species, though nut crops will be much reduced in an exposed location. The foliage is also attractive to grazing animals, if they can get to it. The leaves are very palatable to cattle and have been used as a source of cattle fodder; this may be particularly useful if coppicing is undertaken in summer.

Hazel leaves are relatively high in nitrogen, containing, on average, 2.2 per cent nitrogen as well as 0.7 per cent potassium and 0.12 per cent phosphorus. Hazel leaf litter is sufficiently rich in nitrogen and potassium to benefit other nearby crops as a green manure crop.

Both hazelnuts and filberts support many types of wildlife (though some will eat the nuts!), including insects, birds and mammals; they are highly valued in conservation work. Bees (particularly bumblebee species) are attracted by the early pollen, which can be of considerable benefit to them and they may in turn increase nut set.

Hazelnut female flowers are tiny.

Red catkins on 'Red Filbert'.

Cultivation

Hazelnuts crop best in areas with cool, moist summers and mild to cool winters – areas with oceanic-influenced climates. Winter temperatures below -10°C (14°F) during flowering in the winter may damage the male flowers. Hazels have a similar winter chill requirement to apples.

Hazels are fairly shade-tolerant, often being seen as part of the understorey in forests. However, nut production is severely curtailed in shade, and a position in full sun or only very slight shade is required for good crops. Frost pockets should be avoided because of the early flowering period.

Planting distances vary with the system used. For just a few trees, plant at 4.5-7m (14'6"-23') apart to allow a full, rounded crown to form.

With lines of trees in an orchard-type planting, trees in the row can be closer (3-5m/10-16'), with rows typically 6-9m (20-30') apart, depending on the access needed. Plants don't need staking.

Usually plant a mixture of cultivars for good cross-pollination. Most cultivars will cross-pollinate, though there are incompatibilities (see 'Cultivars', page 159). Ideally, the pollenizing cultivar will mature its crop at the same time as the main crop, and be pollinated by it too.

Rootstocks and soils

Most hazelnut cultivars are grown on their own roots. Because hazel is a suckering species, grafting on to seedling hazel rootstocks risks having rootstock suckers supplant the cultivar in time.

Hazel likes a well-drained soil that doesn't dry out too much. A wide range of soil pH is tolerated (pH 5.5 to 7.5), the optimum being around 6.0. Soil of poor to moderate fertility is most suitable – a very fertile soil may result in excess vegetative growth at the expense of cropping.

Pollination

Trees flower in late winter, with cross-pollination essential. Severe frosts and wet and windy conditions damage flowers; rain clears the air of pollen grains and excess moisture destroys pollen viability (which even in sunny weather is only for a few hours). Commercial plantings use a pollinating cultivar every third row or so, thus you should ensure a pollinator is within about 15m (50') of each cultivar (can be further in a dry climate). If you have wild hazel hedges nearby, these can be a good source of pollen.

Pollen is released in short bursts during relatively dry periods. Although pollen germinates as soon as it reaches the receptive female flower, fertilization doesn't actually take place until some 4-5 months later, in midsummer. This could be why hazel bears a high percentage (up to 30 per cent in some cultivars) of empty nuts, because sometimes the pollen dies before fertilization. Following fertilization, female flowers develop into nuts extremely rapidly, with 90 per cent of growth within 4-6 weeks.

Feeding and irrigation

Hazelnuts are not hungry, and in commercial orchards little if any nitrogen is added. On poor soils, heavily cropping trees may need extra nitrogen and potassium. Adequate levels of boron are also required for good crops.

There is good potential for using hazels in agroforestry systems. Traditional European systems used for many years include a silvopastoral system, with sheep grazing the pasture beneath hazelnut orchards; and interplanted systems of hazels and vines. In Kentish orchards, gooseberries and currants were traditionally interplanted with young hazels. Pasturing with sheep has the added benefit of controlling unwanted sucker growth.

In Mediterranean and dry climates, irrigation is essential in summer, and most commercial orchards use drip systems or micro-sprinklers. In moister climates, mulching of trees should ensure irrigation is unnecessary.

Pruning

Hazel is naturally a multistemmed suckering shrub, and most traditional and low-input systems retain the multistemmed form. The number of main trunks may be limited to five or so – more in some systems – by pruning every few years. But excess suckers and crossing branches are still removed annually.

Female flowers are produced from buds of the past season's growth (similar to peach and almond), so new growth must be maintained for a good crop. A good nut crop requires a large bearing surface (like most nuts – the bearing surface is the surface area of the tree canopy) and 15-22cm (6-9") of new growth per year. It is the weaker horizontal shoots that carry most of the nut crop – vigorous vertical shoots carry very few nuts.

Hazel is also well known as a coppice species, but after coppicing there will be no nut crops for a year or two, before crops start again on the growing bush. Coppicing to keep trees small (e.g. in a small garden) may be feasible, but regrowth is fast and you might find you need to coppice again after five years, just as crops are starting to build again.

Modern commercial systems want to machine-harvest and thus get access closer in to the trunks. Trees in these systems are usually trained with a single short trunk, and the top growth is regularly pruned to keep it to a vase or goblet shape with 8-12 main branches Many suckers are produced annually and these must be removed. Some commercial systems use a hedge system, where lines of trees are trimmed (mechanically or otherwise) on alternate sides every few years.

In France and Oregon, hazelnuts are usually planted at 5 x 3m (16 x 10') spacing, sometimes 6 x 3m (20 x 10') or 5 x 4m (16 x 13'), depending on soil fertility.

Italian training systems include:

* The ceppaia system in Campania, with 4-5 trunks emerging from the ground and with 300-400 trees/ha (120-160/acre).
* In Spain, traditional training consists of a multistemmed bush with 4 or more stems per tree, according to natural growth habit. In new orchards, trees are trained in a single stem vase to facilitate mechanical harvesting and sucker control; tree spacing ranges from 6 x 3m (20 x 10') to 7 x 4m (23 x 13') – 350-550 trees/ha (140-220 trees/acre).
* In England, 'Kentish cobnuts' are a speciality grown on a small scale. These are often unripe nuts hand picked from trees when still 'milky'. Trees are trained with a handful of stems growing about 2m (6'6") high to allow easy access for picking.

Pests

Grey squirrels are the worst pest: they may eat the whole crop before it has even ripened. If you are unable to reduce squirrel numbers, then think twice before trying to grow hazelnuts! Other wildlife pests include mice and crows, and if these become a problem, control measures may be necessary. In North America, blue jays are sometimes a serious pest. Hazels are also susceptible to browsing by deer and cattle, especially when young.

Hazelnut weevil (*Curculio nanum*, syn. *Balaninus nucum*) destroys maturing nuts. These small brown beetles lay their eggs in immature nuts, and the eggs hatch into small white maggots, which feed on the kernel. Many of these infested nuts fall from the tree in late summer, then the maggots bore holes in the shell to escape and pupate in the soil. Infestations are rarely serious on smaller plantings, and the best control measures are to regularly collect and burn early fallen nuts in late summer. You can also hoe or shallow cultivate beneath bushes as early nuts fall, which helps to destroy overwintering pupae. Running chickens below trees can achieve the same aim.

Big bud mite (syn. hazelnut gall mite / filbert bud mite – *Phytoptus avellanae*) is a mite that feeds inside buds, enlarging them (similar to big bud in blackcurrants). Hazels are fairly tolerant and the problem is rarely severe. It is most evident in spring. Heavily galled branches should be pruned out in late winter.

Aphids can feed on leaves and leave sooty moulds on leaves. Damage is usually insignificant and avoided by a diverse planting including predator attractors.

Filbertworm moth.

Symptoms of eastern filbert blight on hazel twigs.

Filbertworm (*Cydia latiferreana*) is a moth whose larvae are significant pests in North America. They lay eggs on developing nuts, the kernels are consumed, and exit holes appear in nuts as the larvae escape to pupate. Pheromone traps trap and monitor numbers. Thicker-shelled cultivars are less likely to be attacked.

Diseases

Hazelnuts are relatively disease-free in Europe and Australia.

Bacterial blight (*Xanthomonas arboricola* pv. *corylina*) is a bacterium, closely related to that causing walnut blight, which can cause a bacterial blight on hazelnuts and filberts. The bacterium causes leaf spotting, dieback of branches and even death of trees in nurseries and young plantations. Trees under stress are most susceptible. Control is by copper sprays.

Eastern filbert blight (*Anisogramma anomala*) is a serious disease in North America. The American hazel, *Corylus americana*, is the natural host of this disease but it is far more devastating to the European species grown in North America. It is not present in Europe. In infected trees, yields rapidly decline and large cankers

form, which can girdle entire limbs. Plants are killed a few years after the cankers spread to the main trunk. Most older cultivars are susceptible, though 'Barcelona' and 'Hall's Giant' are somewhat tolerant, and 'Gasaway' is very resistant. Resistant cultivars have and continue to be bred in Oregon. The best control is to cut out and burn infected branches.

There are various other minor leaf and nut fungal diseases, including mildews, anthracnose, brown rot and grey mould.

Harvesting and yields

Most cultivars start bearing fruit after 3-4 years, with nuts borne on the previous year's wood. Hazels tend to bear erratically and often bear crops biennially.

Just prior to harvest, the orchard floor must be mown short to enable efficient harvest of the

Yields from mature hazelnut trees			
Age	Yield per tree	Yield per hectare	Yield per acre
10-15 years plus	4kg (9lb)	1-2t (1,000-2,000lb)	400-800kg (880-1,760lb)

fallen nuts (many commercial growers in Europe and elsewhere herbicide the orchard floor before harvest).

The husks start turning yellow and the nuts fall off the tree when ripe. Filberts often fall still enclosed in their papery bracts. Small-scale harvest is easy with a Nut Wizard harvester, or you can place a net or tarp under/around trees to collect the nuts. It's worth shaking the trees by hand before you harvest, to bring down any loose nuts. Try to harvest every 1-2 days to minimize losses to mice and squirrels. Nut drop for most cultivars is naturally over several weeks.

Commercial hazel orchards are usually machine-harvested. Sometimes machines are used to shake nuts from trees. Then sweeping machines gather the nuts and, where conditions allow, vacuum harvesters collect them. Next, debris has to be separated off using fans. Two or three passes are required over the ripening season to pick up nuts.

Trees reach maturity at about 10-15 years. Yield per tree can reach 11kg (25lb) in a heavy cropping year; however, average yields from mature trees are listed in the table above. Newer cultivars have the potential to yield considerably more – up to an average of 3t/ha (1.25t/acre) and twice that in a good cropping year. These are in-shell yields; kernel yields are about half these figures.

Processing and storage

Hazel produces a larger percentage of empty nuts than most nut crops. These empty nuts should first be removed by placing the nuts in water – the empties will float.

Surface-dry nuts can be stored in a cool place for a few months. Otherwise the harvested nuts require drying if they are to be stored for more than a few months. Unless grown in a climate with warm, dry autumn weather, it isn't enough to dry in the open air, and drying machines are used. Filberts can be dried with bracts intact.

Shelling is best using an electric cracker; although manual processing is possible.

Nuts in-shell store for 2-3 years, with the best results at low temperatures. Shelled kernels will store for rather less, even at low temperatures. Fresh (undried) nuts store in moist sand for 4 months.

Propagation

Stool layering is used to propagate most cultivars and leads to one-year rooted plants (see Chapter 2, page 62).

In the autumn, trees can be propagated by simple layering – pegging down a low branch, wounding the branch where pegged and covering with soil. After a year the rooted layer may be removed.

The growth and formation of nuts varies a lot. Some are borne singly, others in clusters. Filbert types have leafy bracts covering the nuts; hazel types have open bracts. Here (clockwise from top right) are the young nuts of 'Corabel', 'Hall's Giant', 'Webb's Prize Cob', 'Cosford' and 'White Filbert'.

Softwood and hardwood cuttings can also be taken, though this isn't easy (see Chapter 2, page 59).

Cultivars

Many hazel varieties originated in Spain, Italy, France and Germany, and these in particular have had ~~...~~ in English-spea ~~...~~ common cul ~~...~~

Butler

Cosford

Ennis

Corabel

Hazel synonyms	
Synonym	**Accepted name**
'Avelline Blanche'	'White Filbert'
'Des Anglais'	'Daviana'
'Du Chilly'	'Kentish Cob'
'Duchess of Edinburgh'	'Daviana'
'Fertile de Coutard'	'Barcelona'
'Géant de Halle'	'Hall's Giant'
'Halle Giant'	'Hall's Giant'
'Halle'sche Riesennuss'	'Hall's Giant'
'Lambert's Filbert'	'Kentish Cob'
'Merveille de Bollwiller'	'Hall's Giant'
'Nottingham Cob'	'Pearson's Prolific'
'Red Avaline'	'Red Filbert'
'Red-leaved Filbert'	'Red Filbert'
'Red Zellernut'	'Red Filbert'
'Rote Zellernuss'	'Red Filbert'
'Spanish White'	'White Filbert'
'Webbs Preisnuss'	'Webb's Prize Cob'
'White Aveline'	'White Filbert'
'White Spanish Filbert'	'White Filbert'

Gunslebert

Hall's Giant

Kentish Cob

Lang Tidlig Zeller

Pauetet

Pearson's Prolific

Tonda di Giffoni

Webb's Prize Cob

White Filbert

Commercial hazelnut growing

Hazelnut is grown commercially in many parts of the world. In Britain, the acreage of hazelnut plantations (plats) has fallen from some 7,000 at the start of the century to around 200 acres now, mostly in Kent. The main cultivar in the old plantings is 'Kentish Cob', but more recently other cultivars include 'Butler', 'Ennis' and 'Gunslebert'. Other cultivars worth growing in Britain include 'Corabel', 'Gustav's Zeller', 'Hall's Giant', 'Lang Tidlig Zeller', 'Tonda di Giffoni' and 'Willamette'.

In France, several thousand hectares of hazelnut orchards have been planted in recent decades, most of which are located in the south-west of the country (Aquitaine and Mid-Pyrenees regions – between Bordeaux and Toulouse). The average orchard size is around 14ha (36 acres). Traditional cultivars still used include 'Fertile de Coutard' (synonym of 'Barcelona') and 'Segorbe', while newer cultivars include 'Butler', 'Corabel', 'Ennis', 'Feriale', 'Lewis' and 'Pauetet', with 'Daviana', 'Hall's Giant' and 'Jemtegaard 5' used as pollinators as 8-12 per cent of orchard trees. French hazelnut orchards are quite large (7-10ha/17-25 acre) in flat areas with good soils and rainfall of 700-800mm (28-32"). Tree densities are high (650-800 trees/ha or 260-320 trees/acre) and the operations highly mechanized, with drip irrigation often used.

In the Netherlands there are a few commercial plantings, using 'Butler', 'Cosford', 'Gunslebert', 'Lang Tidlig Zeller' and the Dutch 'EMOA' varieties, which perform especially well in organic systems.

In Denmark, plantings of some 3-4ha (10 acres) exist, with perhaps a total area of 40ha (100 acres). The main cultivars used are 'Kentish Cob' and 'Lang Tidlig Zeller' grown for the 'green' nut market.

In Germany, there are a few commercial orchards, though there is considerable activity by hobbyists. The main two cultivars used are 'Hall's Giant' and 'Kentish Cob'.

In the USA, Oregon is the main area of production. Most traditional orchards are grown as low-input systems, many by part-time farmers. Many orchards have been replanted over the last few decades using cultivars resistant to eastern filbert blight, including 'Barcelona', 'Clark' and 'Lewis'. Oregon State University is continuing to breed resistant varieties. Philip Rutter at Badgersett Research Corporation (see Resources) is also breeding hardy hybrid hazelnut populations for colder parts of North America.

Choosing hazelnut cultivars depends on your aim and, of course, what is available. For a mechanized harvest, cultivars that drop freely from husks are important; with hand-harvesting this isn't important – dwarf or low-vigour cultivars are better for hand-picking. If the aim is to sell shelled nuts, which are usually blanched to remove the bitter pellicle (seed coat), then a high rate of kernel skin removal on blanching is desirable. The most reliable cultivars will be those flowering mid- or late season, which miss the wettest of the winter weather.

When choosing cultivars, at least two are needed for good pollination, preferably more, although wild hazel hedges are good pollinators. Try to include cultivars where the male flowering times of one is the same as the female flowering times of another.

Some nuts are better filled than others, i.e. the percentage of kernel (as compared with whole nut) is larger. Kernel percentages vary from 42 to 50 per cent, depending on the cultivar.

Cultivars best suited to shelled nut production, which have easily removed pellicles when blanched, include 'Casina', 'Corabel', 'Gustav's Zeller', 'Hall's Giant', 'Mortarella', 'Pauetet', 'Riccia di Talanico', 'Tonda di Giffoni' and 'Willamette'.

Key to ripening and flowering times

Ripening season

Each of the following corresponds roughly with a 5-day start period through early autumn:

E = early
EM = early–mid
M = mid
ML = mid–late
L = late

Male and female flowering

Each of the following corresponds to a 10-day start period, so altogether covering a 2-month period in late winter:

VE = very early
E = early
EM = early–mid
M = mid
ML = mid–late
L = late

Closely related cultivars can have incompatible pollen – particularly relevant with recently bred cultivars. If using all recent cultivars, then check for genetic compatibility, for example on the Oregon State University (OSU) hazel publications pages (see Resources).

EFB = eastern filbert blight.

Recommended hazelnut cultivars

Cultivar	Origin	Season	Male flowers	Female flowers	Description
'Barcelona'	Spain	ML	EM	M	(syn. 'Fertile de Coutard') Nuts large (3.6g/0.13oz), round. Vigorous, upright tree, moderately heavy cropping, resistant to bud mites. Old variety still widely used in Europe and the USA.
'Butler'	USA	EM	EM	ML	Nuts large (3.2g/0.11oz), kernels light brown, 48% kernel, oval, sweet, medium-thick shell. Kernel blanching poor. Short husk – most nuts come free easily from husk. Tree vigorous, resistant to bacterial blight, leafs out mid-season, heavy cropper, susceptible to nut weevils. Male flowers abundant.
'Camponica'	Italy	EM	EM	EM	Nuts large (3.7g/0.13oz), round, good flavour. Heavy-cropping tree.
'Casina'	N. Spain	M	M	ML	Nuts small–medium (2g/0.07oz), good flavour, kernels light brown, good kernel %, round with dark bands, thin-shelled. Kernel blanching good. Medium–long husk – medium numbers of nuts fall free. Tree vigorous, spreading, good cropper. Male flowers abundant.
'Clark'	USA	EM	EM	M	Nuts small–medium (2.6g/0.09oz), round, good kernel %. Small trees of moderate vigour with heavy crops. Has some resistance to EFB.
'Corabel'	France	ML	ML	ML	Nuts large (3.7g/0.13oz), sweet, excellent flavour, easily shelled, perfect for blanching. Tree vigorous, upright, heavy and regular yielding, late to leaf out. Male flowers abundant. Recent cultivar.
'Cosford'	UK	M	M	ML	Nuts medium (2.5g/0.08oz), oval, kernels tan, thin-shelled, excellent flavour. Medium-length husk. Tree vigorous, upright, low yielding, late to leaf out.
'Daviana'	UK	EM	ML	ML	Nuts medium–large (3.2g/0.11oz), oblong. Tree vigorous, upright, low yielding, susceptible to bud mites but can be a useful pollinator.
'Delta'	USA	ML	ML		Compact tree. Pollinating variety, highly resistant to EFB and bud mites. Recent.
'Dorris'	USA	ML	EM	EM	Nuts large (3.4g/0.11oz), round. Very recent variety. Nuts medium size, excellent quality. Trees of moderate vigour, spreading, highly resistant to EFB.
'EMOA' 1, 2, 3	Netherlands				Heavy yielding with good-quality nuts and good pellicle removal. Resistant to diseases.
'Ennis'	USA	ML	M	ML	Nuts very large (3.8g/0.13oz), kernels light brown, 46% kernel, round, excellent flavour, do not blanch well. Medium-length husk, moderate fall free of husk. Tree vigorous, high yielding, mid–late leafing, resistant to bud mites, susceptible to botrytis. Male flowers abundant.

Recommended hazelnut cultivars

Cultivar	Origin	Season	Male flowers	Female flowers	Description
'Epsilon'	USA	M	L		Compact tree. Pollinating variety, highly resistant to EFB. Recent.
'Eta'	USA	M	M	ML	Nuts medium (2.8g/0.09oz), round. Tree of moderate vigour, upright, low yielding. Pollinating variety, highly resistant to EFB. Recent.
'Felix'	USA	M	ML	M	Nuts medium (2.5g/0.08oz), round. Tree vigorous, upright, low yielding. Pollinating variety, highly resistant to EFB. Very recent.
'Feriale'	France	EM	EM	L	Nuts large, good sweet flavour. Tree precocious (early flowering), a 'Butler' cross, recent variety.
'Gamma'	USA	M	M	ML	Nuts small (2.5g/0.08oz), round, good for blanching. Tree vigorous, upright and spreading, moderate yielding, pollinating variety, highly resistant to EFB. Recent.
'Gem'	USA	L	M	EM	Nuts very large (4.8g/0.17oz), long, very good for blanching. Tree vigorous, upright and spreading, poor cropper.
'Grifoll'	Spain	ML	ML	ML	Small, oblong nuts (2g/0.07oz). Heavy cropper.
'Gunslebert'	Germany	ML	M	M	Nuts medium (2.7g/0.09oz), 45% kernel, oval, good flavour. Small-medium husk length, moderate fall free of husk. Tree moderately vigorous, low yielding, resistant to bacterial blight and bud mites. Male flowers abundant.
'Gustav's Zeller'	Germany	E			Nuts small, medium-thick shell, excellent for blanching. Tree vigorous, upright, low yielding.
'Hall's Giant'	Germany	M	ML	M	(syns. 'Halle'sche Riesennuss', 'Merveille de Bollwiller')
'Jefferson'	USA	ML	ML	ML	Nuts large (3.4g/0.11oz), kernels golden-brown, 40%, oval-round, thick-shelled, good flavour. Medium husk length, moderate kernel blanching. Tree vigorous, upright, low yielding, mid-late leafing, very resistant to bud mites, not susceptible to nut weevils. Male flowers abundant.
'Jemtegaard 5' ('J-5')	USA	L	M	EM	Large nuts (3.7g/0.13oz), round, good for blanching, very high yields. Tree of moderate vigour, upright, highly resistant to EFB. Very recent.
'Kent Cob'	France?	M	ML	ML	Nuts medium size (2.3g/0.08oz), 48% kernel, oval, thick-shelled, do not blanch well, excellent flavour. Husk long - few nuts fall free of husk. Tree moderately vigorous, spreading, medium-high yielding, mid-late leafing, resistant to bud mites. Harvested as speciality crop in UK; picked young at milky stage.

Recommended hazelnut cultivars

Cultivar	Origin	Season	Male flowers	Female flowers	Description
'Lang Tidlig Zeller'	Germany	M			(syn. 'Lange Landsberger') Nuts small–medium size, kernel golden-brown, thick-shelled. Tree moderately vigorous, spreading, leggy.
'Lewis'	USA	M	EM	M	Crops over a long period, good-quality nuts. Tree moderately vigorous, upright, heavy cropper. Has some resistance to EFB. Recent variety.
'McDonald'	USA	E		M	Tree very open, bears good crops of large nuts, trees resistant to EFB. Very recent.
'Morell'	Spain	ML	M	EM	Vigorous tree.
'Mortarella'	Italy	EM	M	M	Nuts large (2.8 g/0.09oz), 50% kernel, oval, blanches well, good aroma. Husk length medium. Tree moderately vigorous, upright, high yielding, resistant to bud mites. Male flowers abundant.
'Negret'	Spain	ML	EM	EM	Nuts medium size (2.4g/0.08oz), oval, thin-shelled, good kernel %. Tree of low vigour, heavy cropping, susceptible to bud mites.
'Nocchione'	Italy	EM	E	M	Low kernel %. Used as a pollinizer.
'Pauetet'	Spain	ML	EM	M	Nuts small–medium size, 47% kernel, excellent for blanching. Tree vigorous, heavy yielding.
'Pearson's Prolific'	UK	M	ML	M	(syn. 'Nottingham Cob') Nuts medium (2.2g/0.07oz), round, thick-shelled, downy, good flavour. Tree compact, moderate yielding, regular bearer. Male flowers abundant.
'Red Filbert'	Germany	M			(syns. 'Red Zellernut', 'Red Zellernuss') Nuts large with red husks. Vigorous, upright tree with reddish-bronze leaves.
'Ribet'	Spain	ML	E	M	Nuts medium size (2.2g/0.07oz), roundish-oval. Tree vigorous, spreading, good cropper.
'Riccia di Talanico'	Italy	EM	ML	ML	Nuts large (2.8g/0.09oz), oval-round, thin-shelled, kernel light brown, 53%, good flavour, excellent for blanching. Husk length medium, few nuts fall free of husk. Tree vigorous, moderate–high yielding, leafs out early to mid-season. Male flowers abundant.
'Sacajawea'	USA	E	EM	E	Nuts medium (2.8g/0.09oz), round, excellent quality. Trees compact, resistant to EFB. Recent variety.
'San Giovanni'	Italy	EM	E	E	Nuts medium size (2.5g/0.08oz), oblong. Long nuts, good pellicle removal. Heavy cropper.
'Santiam'	USA	EM	EM	L	Moderate-quality nuts. Trees resistant to EFB. Recent variety.
'Segorbe'	France	ML	EM	M	Nuts large (2.8g/0.09oz). Tree vigorous, heavy cropper, susceptible to bud mites.

Recommended hazelnut cultivars

Cultivar	Origin	Season	Male flowers	Female flowers	Description
'Theta'	USA	ML	L	M	Nuts small (2.3g/0.08oz), round. Tree vigorous, upright, low yielding. Pollinating variety, highly resistant to EFB. Recent.
'Tonda di Giffoni'	Italy	ML	EM	M	Nuts large (3.1g/0.10oz), round, thin-shelled, kernel light brown, 48%, excellent for blanching, excellent flavour. Husk length medium. Tree of low vigour, upright, moderate to high yielding, very adaptable, resistant to bud mites, resistant to EFB. Male flowers abundant.
'Tonda Gentile delle Langhe'	Italy	E	E	M	Nuts medium size (2.6g/0.09oz), round. Very adaptable and excellent nut quality. Vigorous tree, moderate yielding, susceptible to bud mites.
'Tonda Gentile Romana'	Italy	E	E	M	Large (3g/0.10oz) round nuts, good aroma, moderate pellicle removal. Tree of medium vigour, high yielding, very resistant to bud mites.
'Tonda Rossa'	Italy	L			Round nuts, good flavour. Tree a poor cropper.
'Webb's Prize Cob'	UK	EM			Nuts large (2.9g/0.10oz), oval, kernels light brown. Tree moderately vigorous, spreading. Male flowers abundant.
'Wepster'	USA	M	ML	M	Nuts small (2.4g/0.08oz), round. Tree vigorous, upright, heavy cropper. Highly resistant to EFB. Very recent.
'White Filbert'	France?	M	EM	EM	Nuts medium (2.4g/0.09oz), thin-shelled, kernel 50%, excellent for blanching, excellent flavour, high oil content. Moderate number of nuts fall free of husk. Tree heavy yielding, resistant to bud mites. Susceptible to weevils; poor cropper.
'Willamette'	USA	ML	E/EM	E	Nuts medium size (2.8g/0.08oz), light brown, long and pointed, white-skinned, excellent flavour. Tree vigorous, upright and spreading, heavy and regular yielding.
'Yamhill'	USA	M	E	M	Nuts small (2.3g/0.08oz), round, good for blanching, good kernel %. Small, spreading, productive trees bearing small nuts. Highly resistant to EFB. Very recent.
'York'	USA	M	M	M	Nuts medium (2.8g/0.09oz), round, good for blanching. Tree of moderate vigour, spreading, moderate yielding. Pollinating variety, highly resistant to EFB. Very recent.
'Zeta'	USA	M		ML	Compact tree. Pollinating variety, highly resistant to EFB. Recent.

Where no details are given, this is because information is not available for these cultivars.

HEARTNUT

(*Juglans ailantifolia* var. *cordiformis*)

ZONE 4-5, H7

Long used in its native range (Japan, but no longer known in the wild) as a nut tree, the heartnut was introduced to North America in the 1800s. Useful for its hardiness (hardier than walnut), several selections were made from seedling stands through the 20th century, especially by members of the excellent Northern Nut Growers Association (NNGA).

Heartnut tree.

Female heartnut flowers.

Male heartnut flowers.

Work in recent decades has improved cracking quality (a problem for a long time), and further improved selections have been chosen – notably by nurserymen and amateur plant breeders Ernie Grimo and Doug Campbell in Ontario, Canada. Heartnuts now have good potential as commercial and garden nut trees; they also make striking and beautiful trees with large, spreading branches and huge tropical-looking leaves. Heartnuts are fast-growing, medium or large deciduous trees reaching 15m (50') or more high, with a 15m (50') spread. A broad, rounded crown forms quite a dense canopy. Light grey bark has dark vertical cracks as trees age, and the young branches and leaf stalks are covered with dense glandular aromatic hairs. Heartnut trees live for around 100 years.

The compound leaves are huge: 40-100cm (16-39") long, with 11-17 oblong or elliptic leaflets, each 7-15cm (3-6") long by 3-4cm (1⅛-1⅝") wide, densely serrated. They are downy on both surfaces, especially beneath.

The male flowers are yellowish-green catkins about 10-30cm (4-12") long. Female flowers are small – similar to walnut, except they are borne on vertical flowering stalks in clusters of 10-20. The flowering stalks droop as the nuts grow, eventually hanging downwards. Flowering is in late spring, usually before walnut, but they are much more frost-resistant than walnut flowers.

The nuts, enclosed in green roundish-conical husks, are formed in clusters of 10-20. The husks are downy, with glandular sticky aromatic hairs.

Nuts of improved cultivars are thin-shelled, 2.5-3.5cm (1-1⅜"), flattened, smooth and pointed with a broad heart-shaped base. They are quite different from the *Juglans ailantifolia* (syn. *sieboldiana* – Japanese walnut), which has nuts like English walnuts but with a point at the top.

Uses

Heartnut kernels are every bit a walnut in flavour and quality. They have a gentle walnut flavour, without the bitter aftertaste of fresh walnuts – in taste tests, a majority of people preferred the taste to that of walnuts. They can be used in place of walnuts in all situations – eating raw, cooked, in baked goods, confectionery, ice cream, etc.

The nuts contain up to 60 per cent of oil. Heartnut oil is delicious and can be used as a salad dressing or cooking oil. Its nutritional content is almost identical to walnut (see Appendix 1), though heartnut contains lower phenolic content, which may explain the 'sweeter' flavour. The pellicle of heartnut kernels is not bitter, as it is in most walnuts.

Like walnuts, the nuts (with husks) can be pickled in the soft green stage before the shells harden (usually in July) and are then very rich in Vitamin C. At this stage the whole nuts, with husks, can be blended whole and be sweetened with honey, then used as a nut marmalade.

Secondary uses of heartnut

Traditionally the bark has been used medicinally, for example as an astringent, diuretic and a kidney tonic. Both the bark and nut husks are rich in tannins and can be used for dyeing a brown colour without the use of a mordant.

The wood is used in Japan for cabinet making, utensils and gunstocks. It is dark brown, light, soft, not strong and not easily cracked or warped.

The shells burn well.

Cultivation

Heartnuts are well suited to temperate maritime regions with regular summer rainfall; unlike most walnut family members, they grow and crop better in cooler regions than warmer ones.

Although trees tolerate wind exposure and tend to have strong branches attached a wide angles to the trunk, they will crop much less in an exposed position due to poorer pollination.

Heartnuts are large, spreading trees. They ultimately need 10-15m (33-50') of space to develop a large, rounded crown. However, it is quite possible to plant trees at half this spacing, with the aim of thinning trees out after 10 years or so. Alternatively, heartnut trees can be planted at final spacings, with other fruit trees or productive trees/bushes interplanted, to produce crops for 7-10 years before they are shaded out.

Like other members of the walnut family, heartnut trees produce juglone from roots and leaves. This has an allelopathic effect on some other plants, notably apples and white pines, suppressing their growth and even killing them. It would be wise to make sure those species are not planted too close.

Rootstocks and soils

Heartnuts are grown either as seedling trees on their own roots, or as grafted trees using heartnut or walnut seedling rootstocks. Most soils of reasonably good fertility are fine. Heartnuts like a moist, well-drained soil and full sun.

Pollination

Some heartnut trees are at least partially self-fertile, but many are not, so two or more different selections should be grown for cross-pollination. Flowering is around late spring, the male and female flowers ripening at slightly different times. The flowers are wind-pollinated. Adequate pollination is also dependent on good weather conditions: a rainy period can lessen crop yield.

Flower initiation also depends on suitable conditions during the previous summer. A warm, dryish summer is likely to lead to more flowers the following spring.

Flowering overlaps with that of butternuts, and cross-pollination between species does occur (the offspring being buartnuts). Cross-pollination with walnut is rarer, but it can occur.

Feeding and irrigation

Little feeding is needed, as heartnuts are not very heavy croppers.

Pruning

Little pruning is needed. Low branches should be pruned off as the tree grows to give a clean trunk, allowing access beneath trees for harvesting. Growth is fast, up to 1m (3') per year.

The nuts develop in roundish-conical husks.

Pests

Squirrels and crows are the notable pests. Heartnuts are resistant to walnut husk flies (*Rhagoletis completa*).

Diseases

There are no serious diseases of note. Heartnuts are resistant to walnut blight (*Xanthomonas campestris* pv. *juglandis*), leaf spot or walnut anthracnose (*Gnomonia leptostyla*), and butternut canker (*Sirococcus clavigignenti-juglandacearum*). In warm climates they can be susceptible to bunch or broom disease – a mycoplasma-like organism that causes dense clumps of branches ('witches' brooms') to form on limbs.

Harvesting and yields

Grafted trees start cropping about 5 years from planting, seedlings 6-8 years from planting. Crops then increase rapidly to a maximum at about 15-20 years of age. Yields from grafted cultivars reach an average of about 30-35kg (66-75lb) per tree. Yields of 5 t/ha (2t/acre) or more are achievable. These are in-shell yields; kernel yields are about one-third of these figures.

Heartnut tends to be a reliable cropper, rarely taking years 'off'. Trees can remain highly productive for 75 years or more.

Nuts fall within the green husks, which soften at ripening, around mid-autumn. They can be harvested using the same equipment as used for walnuts.

Processing and storage

The husks must be removed promptly from around the nuts – if they rot and blacken, then within a few days the nut and kernel can be affected and ruined.

Heartnut husks soften as the nuts fall and the nuts are fairly easy to remove. For a few nuts, just wear gloves and remove the husks by rubbing – washing the nuts helps remove any remaining husk pieces.

On a larger scale, you can use a concrete mixer to remove husks: place the crop with some stones and water in the mixer, and run it for a few minutes. Then pour the mixture out and remove the nuts (most of the husk pieces float; the nuts sink in water).

Drying is essential for long-term storage. Like other walnut family nuts, it is important not to dry at too high a temperature, as the oils in the nuts may go rancid – about 40°C (104°F) is best.

Blow warm air through layers of nuts in a dehydrator or drying machine.

Dried heartnuts will store for 3-4 years or so. With improved cultivars, the nuts crack cleanly along the seam (suture) into two halves. The inner side of the nut is smooth and the kernels can usually be extracted whole or at least in halves.

Propagation

Seedling propagation is quite easy – seeds require a few months of cold stratification, so either sow in autumn and protect against rodents, or stratify in a fridge and sow in spring. One-year seedlings grow 30-70cm (12-28") tall. Heartnut seedlings resent transplanting so best to grow them in a deep cell container. Seedlings will not come true (i.e. they will not be identical to the mother tree), and depending on the pollinator may be heartnut, Japanese walnut or buartnut type and shape.

Propagation of cultivars is usually by grafting. Grafting is quite difficult and needs warm temperatures around the graft union – in spring, the use of a hot grafting pipe can be useful. The rootstocks used are seedlings of heartnut, butternut, black walnut or walnut. Chip budding and greenwood (semi-ripe) tip-grafting in early summer can also be successful.

Established trees can be top-worked by spring bark grafting, chip budding and cleft grafting. R. J. Garner's *The Grafter's Handbook* (see Resources) contains detailed information on these grafting techniques.

Layering can work if there is a convenient low branch.

Cultivars

When choosing cultivars, ease of nut cracking is a major factor. Good cracking varieties have nuts that fracture reliably on the suture line; internal shell cavities with four smooth, open lobes that easily release the kernel; and the kernel should fall from the shell in one or two pieces. The 'cracking' column in the table on page 170 indicates how good the cracking is.

Flavour and eating quality doesn't differ very much between cultivars. With good cultivars, kernel percentage is usually 30-35 per cent of the whole in-shell nut, and the nut count is 132-176/kg (60-80/lb).

Most of the recent heartnut selections have been made in the Great Lakes region of North America – from Ontario in Canada, and New York and Pennsylvania in the USA. This involved selecting the best varieties from seedlings of the older varieties. Some older selections were made a few decades ago in British Columbia.

For most cultivars the flowering habit is unknown, but ideally match protandrous flowering cultivars (males flowering first) with protogynous cultivars (females first).

Heartnut – Campbell CW3

Heartnut – Fodermaier

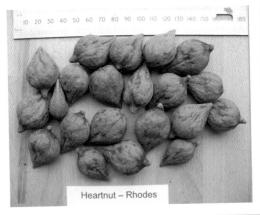
Heartnut – Rhodes

Heartnut Wright

Recommended heartnut cultivars

Cultivar	Origin	Nut size	Cracking	Yield	Comments
'Brock'	USA	Large	Good	Good	Late season. Protogynous.
'Callandar'	USA	Medium	Good	Moderate	
'Campbell CW1'	Ontario	Medium–large	Moderate	Very good	Mid-season. Productive annual cropper. Protogynous.
'Campbell CW2'	Ontario	Medium	Moderate	Moderate	
'Campbell CW3'	Ontario	Medium	Good	Good	Mid-season. Compact tree. Protandrous.
'Campbell CWW'	Ontario	Medium	Moderate	Good	
'Ebert'	USA	Large	Good	Good	
'Etter'	USA	Small	Good	Moderate	
'Fioka'	BC	Medium	Good	Good	Vigorous tree, previously thought to be a buartnut.
'Fodermaier'	USA	Medium	Good	Moderate	Mid- to late season. Particularly hardy tree. Protogynous.
'Frank'	USA	Medium	Good	Moderate	Late leafing.
'Grimo Manchurian'	Ontario	Medium	Good	Good	May be a heartnut x Manchurian walnut.
'Imshu'	Ontario	Medium	Good	Very good	Very hardy and productive tree. Protogynous.
'Locket'	Ontario	Medium	Good	Very good	Protandrous.
'Okandra'	BC	Small	Good	Good	
'Pike'	USA	Medium	Moderate	Moderate	
'Rhodes'	Ontario	Medium	Good	Good	Late season. Self-fertile and late flowering. Protandrous.
'Rival'	USA	Medium	Good	Good	
'Schubert'	USA	Medium	Good	Good	
'Simcoe'	Ontario	Large	Good	Very good	Heavy bearer. Protandrous.
'Stealth'	Ontario	Large	Good	Good	Protandrous. Broad, large nuts.
'Stranger'	USA	Small	Good	Good	Early ripening tree. Protandrous.
'Walters'	USA	Large	Moderate	Moderate	
'Westfield'	USA	Medium	Moderate	Good	
'Wright'	Ontario	Medium	Poor	Good	

HICKORIES

(*Carya* spp.)

ZONE 5-6, H7

The hickories are closely related to walnuts (both are members of the family Juglandaceae) and, like them, are usually large deciduous trees that can live to a great age (400-500 years) and tend to form upright cylindrical crowns when grown in the open.

A nine-year-old shagbark hickory (*C. ovata*) at 2.7m (9') high.

Catkins on mockernut (C. tomentosa).

Most *Carya* species have edible nuts. Pecan (*C. illinoinensis*) is described separately on page 202. The main species of interest here are:

❋ Shagbark hickory (*C. ovata*)
❋ Shellbark hickory (*C. laciniosa*)
❋ Mockernut (*C. tomentosa*)
❋ Hicans – hybrids of pecan (*C. illinoinensis*) and other *Carya* species

Hickories have alternate pinnate leaves, each with 3-17 leaflets. Most hicans have 9 or 11 leaflets, compared with pecan (13), shellbark (7) and shagbark (5).

The bark on hickories such as shagbark (*C. ovata*) comes loose and falls in long strips, hence the name. Hican bark is similar to pecan bark. Hickory foliage is aromatic.

Male flowers are borne on slender, drooping catkins, which arise from lateral buds; female flowers are borne in a spike at the end of the current season's shoot. The latter are followed by large fruit consisting of a single nut surrounded by a leathery skin (husk), which may split open at maturity. Hican fruit tends to be intermediate between the parents.

All species have pronounced taproots, which securely anchor the trees if soil conditions allow. Hickories can easily be confused (especially by leaf) with walnuts, but young hickory shoots have non-chambered pith, and the nuts are smooth-shelled. Most species are quite hardy, though young plants are sometimes damaged by late spring frosts.

Some 14 species of *Carya* are found in eastern North America (plus at least 12 interspecific hybrids), and a further 6 in China. The nuts, which are of comparable sizes to walnuts, are rich in oils and edible from most, though not all,

Young leaves of shagbark hickory (*C. ovata*).

of the species, and the better ones have a rich walnut-like flavour. They tend to be hard-shelled, like black walnuts.

Trees often take 10 years or more to start cropping, which is one limitation of hickory as a commercial nut crop. They are better suited to low-input, sustainable agricultural systems, where the long-lived multifunctional trees are a valuable resource for food, fuel and high-quality timber.

Uses

The edible nuts have a sweet kernel contained within a shell that varies in thickness from species to species and also within a species. Thick-shelled species are difficult to crack and may contain kernels weighing only 20 per cent of the total nut. The kernels can be eaten raw or roasted; they can also be made into a nut butter, an oily 'hickory milk', or ground into a flour and used in breads, etc.

Kernels are rich in oils and resemble walnuts in richness of flavour, and pecan and walnut in nutrient content. See Appendix 1 for a breakdown of their nutritional composition.

A good-quality culinary oil can be extracted from the species with edible nuts.

Secondary uses of hickories

Hickory nut oil has been used as a fuel for oil lamps and medicinally for rheumatism. In China it is used in paint making.

Several (probably all) species can be tapped for the sap, which is concentrated to make a syrup or made into a wine, etc.

The shells are used for making activated charcoal in China.

Hickory wood is well known for its strength and resilience and is excellent for tool handles (hammers, picks, axes, etc.). The heartwood is brown or reddish-brown and sold as 'red hickory'; the sapwood is sold separately as 'white hickory'. Hickory wood is straight-grained, coarse-textured and heavy (820kg/m^3 or 51lb/ft^3). Stiff and shock-resistant, it has high bending and crushing strength and excellent steam-bending properties.

High-quality timber is used for the manufacture of skis, gymnastic bars and other athletic equipment (golf club shafts, lacrosse sticks, tennis racquets, basketball bats, longbows); also as a flooring material for gymnasiums, roller rinks and ballrooms. Some wood is used in furniture making, piano construction, for butcher's blocks, wall panelling and interior trim, dowels, ladder rungs and pallets, heavy sea fishing rods, drumsticks, wheel spokes and vehicle bodies. It makes excellent firewood and charcoal, and is used in cheese and meat smoking. The wood of water hickory (*C. aquatic*), bitternut hickory (*C. cordiformis*) and nutmeg hickory (*C. myristiciformis*) is considered inferior to other hickories.

A yellow dye is obtained by using the bark of most species with an alum mordant. Other dyes are obtained from the leaves and fruit husks.

Bark on mature trees of shagbark hickory (*C. ovata*).

174

Cultivation

All hickories prefer a climate with hot summers, and they need a position in sun but with shelter from strong winds.

Hickories are slow-growing, especially when young, but can eventually become large, spreading trees. They ultimately need about 12-15m (40-50') spacing between trees, but if growing at this spacing it makes sense to interplant with one or two other trees (e.g. fruit trees), which will have 15-25 years to grow and crop before the hickories will need the space.

Hickories are late to leaf out in late spring and relatively early to drop their leaves in mid-autumn. Thus there is good potential for growing an undercrop, particularly one that is cropped in late spring. When in leaf they cast a relatively heavy shade.

Hickories, like walnuts, contain juglone in the leaves (and probably the roots too). This substance can have growth-suppressing effects on some other plants, such as apples and white pines.

Rootstocks and soils

Hickories are mostly grown as seedling trees on their own roots. Grafted trees use hickory species (usually the same as the scion) for r ootstocks.

They prefer a good fertile soil, preferably a deep moisture-retentive loam, though they tolerate both light and heavy soils, and acid and alkaline conditions. Transplanting should be undertaken with care because of the long fleshy taproot: for their first few years, young trees form a taproot with only a few lateral feeder roots, and this taproot is usually longer and thicker than the above-ground stem. If raising plants, either grow them in open-bottomed containers that air-prune the taproot – i.e. roots emerge from the bottom, causing the root tip to die and subsequent branching of the root system in the container. Alternatively, undercut the taproot at least a year before transplanting – i.e. use a blade or spade to cut through the taproot at 20-25cm (8-10") below ground level, forcing them to branch. A branched root system transplants more successfully than a single thick taproot.

Hickories are very slow-growing for the first five years or so, but then make good growth; planting in tree shelters may be advantageous.

Pollination

Although all species are monoecious (bearing both male and female flowers), hickory trees are usually not very self-fertile, with male and female flowers maturing at different times on the same tree (similar to walnuts). Growing a mixture of cultivars (or seedlings) should ensure good pollination.

The flowers are produced in mid-spring. They are wind-pollinated, so trees should be within about 50m (160') of each other for adequate pollination. Different hickory species may also cross-pollinate. You may need windbreaks to ensure strong winds don't just blow the pollen out of the orchard.

Feeding and irrigation

Little of either is needed. Hickories are not heavy producers, but after a good cropping year it is worth feeding trees with something containing nitrogen and potassium.

Trees are deep-rooting and quite drought-tolerant. In very dry years the nut size will reduce with unirrigated trees.

Pruning

Lower branches are usually pruned out in the first few years as the trees grow, to allow for 1.8m (6') or more of access beneath. Apart from that there is little, if any, pruning required.

Pests

Squirrels are the main pest. Crows may also be a nuisance. See 'Pecan' (page 206).

Diseases

None of significance. All hickories are resistant to honey fungus (*Armillaria* spp.). See 'Pecan', page 207.

Harvesting and yields

Harvest is usually in late autumn. The outer husk (outer shell) splits along sutures and either releases the hard-shelled nut or falls still encasing the nut.

Many hickories crop biennially, with crops in 'on' years reaching up to 40kg (88lb) per tree. (As humans, we tend to view biennial bearing as 'unreliable' and a nuisance, and nut/fruit tree breeders try to select trees that are annual in bearing habit. However, from the trees' point of view, biennial bearing makes absolute ecological sense, as it breaks pest cycles on the nuts/fruit. Trees that crop annually almost always have lots more pest problems.)

Husks splitting to release nuts on shagbark hickory (*C. ovata*).

Processing and storage

On some trees, nuts fall freely from the husks; on others they fall within the husks. With the latter, husks must be removed (e.g. by the concrete mixer method, see Chapter 3, page 69). The nuts then need drying for long-term storage. Dried hickory nuts should store for 2-3 years at room temperature.

Propagation

As with walnuts, grafting of hickories is difficult but remains the only way of propagating named selections. Rootstocks used are normally seedlings of the same species (or one of the parents), which make strong graft unions. Grafting on to pecan gives slightly more vigorous trees. Grafts or seedlings should be grown in deep containers that encourage air-pruning of roots.

Seeds of most species require three months of cold stratification before germination will take place.

Cultivars

Grafted trees are generally unavailable in Europe, where seedlings from named cultivars are most commonly planted, giving good prospects of decent bearing trees. Nut size varies, with cultivars of shellbark (*C. laciniosa*) having nuts 25-40mm (1-1⅝") long, mockernut (*C. tomentosa*) 30mm (1⅛") long, and shagbark (*C. ovata*) 25-30mm (1-1⅛") long.

Most hickory seedlings from named varieties start to fruit about 10-12 years from seed; precocious (early fruiting) cultivars start fruiting in 5-7 years. Seedlings from random wild trees can take 20-40 years to crop.

Some 200 cultivars have been selected in the USA over the last 100 years, but many of these are now lost to cultivation.

Shagbark hickory Glover

Shellbark hickory - Henry

Shagbark hickory Neilson

Shagbark hickory Weschcke

Shagbark hickory – Wilcox

Shagbark hickory Yoder

Key to hickory cultivar table

Hican[1] = *C.* x *brownii* (*C. cordiformis* x *C. illinoinensis*)

Hican[2] = *C. illinoinensis* x *C. ovata*

Hican[3] = *C. illinoinensis* x *C. laciniosa*

The best hican varieties are generally accepted to be 'Burton', 'Dooley Seedling' (syn. 'Dooley Burton') and 'James'.

Recommended hickory cultivars

Cultivar	Parentage	Comments
'Abundance'	*C. laciniosa* x *C. ovata*	Nuts medium–large, crack well, 36% kernel. Tree precocious, regular bearer, good in northern locations.
'Brackett'	*C. cordiformis* x *C. ovata*	Nuts large, thin-shelled, high kernel %, good flavour.
'Burton'	Hican[2]	Mid-season. Nuts medium size, crack well, medium kernel %, good flavour, drop free of husk. Tree precocious, annual bearer. Good in northern locations.
'Cedar Rapids'	*C. ovata*	Nuts large, good quality. Tree regular bearing. Good for northern regions.
'CES-1'	*C. laciniosa*	Nuts medium size, good flavour. Tree healthy, good in north.
'CES-24'	*C. laciniosa*	Nuts large, 40% kernel, excellent flavour. Tree precocious, good in north.
'CES-26'	*C. ovata*	Nuts crack well, high kernel %. Very early ripening, good for northern regions.
'De Acer'	*C. cordiformis* x *C. ovata*	Nuts medium size, crack well, good flavour. Tree precocious.
'Dewey Moore'	*C. laciniosa*	Nuts thin-shelled, 33% kernel. Tree a light bearer.
'Doghouse'	*C. laciniosa* x *C.ovata*	Nuts large, crack well, good flavour.
'Dooley'	Hican[2]	Mid-season. Nuts medium–large, well filled, crack fairly well, drop free of husk. Tree crops well, good in northern locations.
'Engeman' (syn. 'Missouri Giant')	*C. laciniosa*	Nuts large, crack well.
'Eureka'	*C. laciniosa*	Nuts medium size, crack well, good flavour. Tree precocious.
'Fayette'	*C. laciniosa*	Nuts medium–large, crack well, 34% kernel. Tree a regular bearer, good in northern locations.
'Fox'	*C. ovata*	Nuts medium size, crack well. Good for northern locations.
'Galloway'	Hican[1]	Sweet nuts.
'Gerardi'	Hican[3]	Nuts large, crack well, very good flavour. Tree very precocious.
'Glover'	*C. ovata*	Nuts small, crack well. Good in northern locations.
'Grainger'	*C. ovata*	Nuts large, crack easily, drop free of husk. Tree a good producer. Very tolerant of alkaline soils.
'Hartmann'	Hican[2]	Nuts medium size, thin-shelled.
'Henke'	Hican[2]	Nuts small-medium size, well filled, crack well, good flavour. Tree precocious, light cropper. Good in north.

Recommended hickory cultivars

Cultivar	Parentage	Comments
'Henning'	*C. laciniosa*	Nuts large. Tree vigorous.
'Henry'	*C. laciniosa*	Late season. Nuts large, well filled, crack well, drop in husk. Tree a regular bearer, good in northern locations.
'Hershey'	Hican[2]	Nuts medium size, well filled.
'Hoffeditz'	*C. laciniosa*	Nuts medium size, crack well, good flavour. Tree a regular bearer.
'Jackson'	Hican[2]	Nuts medium size. Tree a good cropper.
'James'	Hican[3]	Nuts very large, crack easily. Tree precocious.
'Kaskaskia'	*C. laciniosa*	Nuts medium–large, crack well, very good flavour. Tree precocious.
'Keystone'	*C. laciniosa*	Mid-season. Nuts large, very good cracking quality, drop free of husk. Tree a regular bearer.
'Lindauer'	*C. laciniosa*	Nuts large, crack well. Tree productive.
'Neilson'	*C. ovata*	Very early season. Nuts medium–large, moderate quality, drop free of husk. Tree precocious, a regular producer. Good in northern regions.
'Pleas'	Hican[1]	Nuts medium size, slightly astringent, 54% kernel, good in northern regions.
'Porter'	*C. ovata*	Nuts medium size, crack well, 47% kernel. Productive tree.
'Ross'	*C. laciniosa*	Nuts large, excellent cracking, good flavour. Tree precocious.
'Scholl'	*C. laciniosa*	Nuts large, crack well. Tree precocious.
'Selbhers'	*C. laciniosa*	Nuts large. Tree early season ripening, good bearer.
'Silvis 303'	*C. ovata*	Nuts large. Tree a good producer.
'Simpson'	*C. laciniosa*	Nuts medium size. Tree a heavy cropper.
'Totten'	*C. laciniosa*	Nuts very large, 25% kernel. Tree vigorous, healthy, good in northern locations.
'Walters'	*C. ovata*	Nuts very large, crack well.
'Weschcke'	*C. cordiformis* x *C. ovata*	Early season. Nuts medium size, thin-shelled, crack well, 53% kernel, very good flavour, drop free of husk. Tree precocious, annual bearer.
'Wilcox'	*C. ovata*	Nuts medium size, crack well, 41% kernel, very good flavour.
'Yoder #1'	*C. laciniosa* x *C. ovata*	Late season. Nuts medium size, thin-shelled, crack well, 40% kernel, excellent flavour, drop free of husk. Tree precocious, good cropper. Good in northern locations.

MONKEY PUZZLE

(*Araucaria araucana*)

ZONE 6-7, H6

Also known as Chile pine, this tree is so distinctive that almost everyone knows it immediately. Growing eventually into a large evergreen tree, it originates from the far south of South America in Chile and Argentina, where the nuts have been an important crop over many centuries.

Monkey puzzle trees lining an avenue in Bicton, Devon, UK.

Since being introduced into England by botanist Archibald Menzies in the late 18th century (after pocketing the nuts from a dinner laid on in Chile), the tree has been widely planted as an ornamental in western Europe. The name 'monkey puzzle' (with variants used in other languages) apparently arose from a chance remark at a Victorian tree planting ceremony: "It would certainly puzzle a monkey to climb that tree."

Relatives of monkey puzzle include *A. angustifolia*, the Paraná pine, from Brazil, Argentina and Paraguay. This also has excellent edible nuts – larger than those of monkey puzzle – but is only just frost-hardy.

Although very slow-growing in its first few years, trees can reach 15-30m (50-100') high. Monkey puzzle has an erect cylindrical trunk up to 1m (3') in diameter in Europe (wider in its native habitat) that is mostly prickly with either living or dead remains of leaves. The bases of large trees are often buttressed. The bark is grey, wrinkled, and marked with rings formed by old branch scars as well as with remains of leaves. Trees can live for 150-300 years, occasionally more.

The symmetrical pattern of the branching is very distinctive. Uppermost branches are ascending, lower ones pendulous; they are shed after some years as they become shaded by higher branches. Branches are produced in regular tiers of 5-7.

The leaves are dark glossy green, broadly triangular, rigid, hard, leathery and sharp-pointed, and 3-4cm (1⅛-1⅝") long; they are arranged in close-set, overlapping spiral whorls completely hiding the shoot. Leaves remain green for 10-15 years but may persist on the tree long after they turn brown and die. Each terminal bud is hidden by a protecting rosette of immature pale green leaves.

The species is dioecious (male and female flowers are normally borne on separate trees), though very occasionally male and female flowers occur together on one tree. Male flowers are produced on cylindrical catkins forming clusters at shoot tips and are brownish with densely packed scales, 7-12cm (3-5") long, shedding pollen in July.

Female cones are wind-pollinated in early summer; the seeds take two or three seasons to develop and are shed early–mid-autumn of the second or third year. Female cones are striking round head-sized objects, scattered singularly and erect on the upper sides of shoots, 10-17cm (4-7") across. They have numerous spirally ranged, golden-tipped leafy scales, each of which is connected to the inward-pointing seed. Dozens of cones are usually borne, each cone bearing 100-300 seeds in their second or third season.

Cones on a monkey puzzle tree.

The seeds ripen to a light reddish-brown colour and are pointed and roughly conical, 25-45mm (1-1¾") long by 12-18mm (½-¾") wide, often with the leafy scale still attached at the wide end.

In South America, reddish round nodules have been observed on the roots of monkey puzzle, and are believed to host bacteria involved in nitrogen fixation. These nodules haven't been observed elsewhere (nobody has looked), but they may require a species of bacteria not present in soils outside South America.

Uses

Monkey puzzle nuts (also known as 'Chile pine nuts' or *piñones*) are starchy, like chestnuts, and have a thin, leathery, easily peeled shell. See Appendix 1 for their nutritional content.

They can be eaten raw, but are more digestible and have a better flavour and aroma when steamed or roasted – the flavour is a mixture of chestnut and tropical plantain. They can be dried, milled into flour and used to make bread. They can also be used for brewing; a spirit is distilled from them in Chile, where the nuts have at times been an important food staple.

Araucaria araucana

Monkey puzzle nuts.

Secondary uses of monkey puzzle

Cultivation of monkey puzzle as a dual-purpose tree, for nuts and timber, is a possibility in mild temperate areas. The timber is valuable and pine-like, being resinous, straight-grained, durable, light, of medium strength, fragrant and pale yellow. It is used for construction, flooring, joinery and interior carpentry, furniture and cabinet making, masts and paper pulp.

The resin from the trunk is used medicinally in Chile, probably in a similar way to pine resin, which is used as a diuretic, irritant and rubefacient (easing the pain and swelling of arthritic joints).

Cultivation

Monkey puzzle prefers a moist, mild maritime climate – towards the west of countries in western Europe and North America is ideal. Older trees tend to be hardier.

Full sun is required for good growth and cropping. Shelter is not essential, as trees are very wind firm, but growth will be faster in a sheltered position. Growth for the first 3-4 years can be 10cm (4") per year, but after planting out can reach 30-60cm (1-2') per year after 10 years. Height after 10 years is around 2m (6'6").

Plant trees at 4-10m (14-33') apart. Because the species is dioecious (male and female flowers are borne on different plants) and it is impossible to tell the sex of seedlings, a group of plants should be planted. If you only plant two trees, there's a one-in-two chance they are the same sex – so plant more than that!

Rootstocks and soils

Only seedling trees are grown and planted.

Monkey puzzle prefers a deep acid soil that is moist but also well drained.

Pollination

Flowering is in early summer, when the weather is usually fine for wind pollination.

Feeding and irrigation

None usually needed.

Pruning

None usually required. Lower branches tend to self-prune by the time flowering begins (if not then a few lower branches may need removing for access beneath).

Pests

There are none of note. Even squirrels would have a hard time climbing this tree.

Diseases

The tree is susceptible to honey fungus (*Armillaria* spp.).

Monkey puzzle trees.

Monkey puzzle leaves.

Harvesting and yields

Monkey puzzles don't start producing nuts until they are 20-30 years of age – so you need to be patient!

The female cones disintegrate on the tree and the nuts fall freely to the ground in early autumn. The long, thin shape of the nuts makes them less easy to harvest mechanically than rounder nuts, but a hazel-sized Nut Wizard will pick them up fine.

Yields from mature trees can reach 10kg (22lb) or more per tree.

Processing and storage

After surface-drying, the nuts will store in cool fridge conditions for a year or so. Longer-term storage is possible with further drying (like chestnuts), after which the nuts will need rehydrating before cooking.

Propagation

Propagation is by seed. There are around 350 seeds/kg (160/lb), and seeds are not dormant.

Even the trunk of monkey puzzle tree is well defended.

They should be sown in spring in warm conditions. If seed is obtained in autumn, keep cool in a fridge until spring, because autumn-sown seed is quite prone to rotting over the winter. Seedlings resent transplanting and should not be potted up for at least a year – it will not speed up their snail-like growth! Pot up in winter after the first year of growth and keep well supplied with nutrients.

Cultivars

There are no cultivars available. Because flowering is dioecious, trees are usually male or female, though trees with both types of flower have been observed.

OAKS
(*Quercus* spp.)

ZONE 3-9, H3-7

The oaks are a large family of some 600 species of evergreen and deciduous trees and shrubs from the Americas, Europe, Asia and northern Africa; they are found in temperate and subtropical zones (and the tropics at high altitude).

Vallonea oak (*Q. ithaburensis* subsp. *macrolepis*).

Oak leaves are alternate and toothed or lobed. Male flowers are borne in slender pendulous catkins; female flowers are borne singly, in twos or more, in spikes. Flowers are wind-pollinated, and the pollen can cause allergies (hayfever) in some folk. Hybridization between different species is common. The fruit – the acorn – is a single-seeded nut, round or ovoid, and partly to almost wholly enclosed by a cup-shaped scaly 'cupule'. Acorns mature either the autumn after flowering or a year later, depending on the species.

All oaks like summer heat, and tolerate high summer humidity. Those from more continental climates (eastern USA, China, Japan) often require good summer heat to ripen their wood properly, and do not perform so well in cooler summer regions with maritime climates. No oaks show extreme cold tolerance but there are a few species hardy to Zone 3 / H7. Many oak species can live for hundreds of years.

There is a long history of human cultures using acorns as a food source, often as a staple crop, in Europe and Asia. The most recent peoples to use acorns as a major food source were Native Americans, who used them widely well into this century. They are still a regular item of use and commerce in a few places, notably Korea and parts of the Mediterranean region.

American oaks can be divided into two groups: white oaks and black (or red) oaks. White oaks mature their acorns in their first year and have leaves with rounded lobes, without pointed tips; in black oaks, acorns take two years to develop, and leaves have bristles or pointed tips.

Uses

All oaks bear acorns, which are edible. Acorns provide a complete vegetable protein and are high in carbohydrates. They contain 16 amino acids, appreciable amounts of Vitamins A and C, and significant quantities of calcium, magnesium, phosphorus, potassium and sulphur (see Appendix 1).

Most oak species produce acorns that are high in tannins, making them bitter and astringent when raw, hence they need processing to remove these potentially harmful substances (see 'Processing and storage', page 190). However, removal of tannins is easy, and the resultant acorn meal resembles that from other nuts in oiliness and flavour. It can be used in soups, stews, breads and biscuits; or dried and ground into a flour. 'Sweet acorns', with low tannin levels, can be used whole or in halves directly in recipes, for example adding to bread.

Acorn beverages have been made in many places, usually being coffee-like, prepared by roasting and grinding acorns. The quality depends on the acorn and technique – chinkapin oak (*Q. muehlenbergii*) was valued for this purpose in the USA, and English oak (*Q. robur*) was used in Europe during the Second World War. In Turkey, 'racahout', a spiced acorn drink, was traditionally made from acorns of holm oak (*Q. ilex*) well into the 20th century.

The oil content varies widely between different species and can reach 30 per cent of kernel. From high-oil types, the oil can be pressed like other nut oils and is comparable in quality to olive oil: it has been used in north Africa (ballota oak – *Q. ilex* var. *ballota*) and the USA (live oak – *Q. virginiana*) as a cooking oil.

Oaks and mycorrhizal fungi

Oaks form numerous symbiotic associations with mycorrhizal fungi, which can significantly aid their nutrition and health (see Chapter 1, page 27). Edible mycorrhizal species associated with oak include those in the table below.

Edible mycorrhizal fungi associated with oak

Latin name	Common name
Amanita caesarea	Caesar's mushroom
Amanita calyptroderma	Coccora
Boletus aereus	Bronze bolete, dark penny bun
Boletus aestivalis	Summer king bolete
Boletus appendiculatus	Butter bolete
Boletus impolitus	Iodine bolete
Boletus luridus	Lurid bolete
Boletus porosporus	Sepia bolete
Boletus pulverulentus	Inkstain bolete
Boletus queletii	Deceiving bolete
Boletus regius	King bolete
Boletus rubellus	Ruby bolete
Cortinarius praestans*	Goliath webcap
Cortinarius violaceus	Violet webcap
Exsudoporus frostii*	Frost's bolete, apple bolete
Gyroporus castaneus	Chestnut bolete
Hygrophorus arbustivus	
Hygrophorus marzuolus	March woodwax
Lactarius quietus	Oakbug or oak milkcap
Leccinum carpini	
Leccinum crocipodium	Saffron bolete
Leccinum quercinum	Orange oak bolete
Russula atropurpurea	Purple brittlegill, or blackish-purple brittlegill
Russula brunneoviolacea	
Russula xerampelina	Crab brittlegill
Sparassis laminosa	
Tuber brumale	Winter truffle

*These species are easily confused with poisonous species.

Evergreen holm oaks (*Q. ilex*) make dense shelter.

Secondary uses of oaks

Oaks have many uses, notably as timber, and as valuable food and shelter for wildlife.

Cultivation

Oaks generally prefer a sheltered site, though a few tolerate maritime exposure. Full sun is required for maximum nut crops, and warm, dry summers tend to favour heavy crops. Most species tolerate moderate side shade and a degree of exposure. Many species develop deep taproots and are drought-resistant when estab-lished, but good acorn production requires reasonably fertile soil and sufficient water.

Planting distances depend on species, ranging from 4m to 15m (13-50'). The larger spacing usually allows room for 10-20 years of inter-planting between trees.

Scattered oak trees have long been common in arable fields and grassland in Europe and North America, though are now less common in arable fields. Because the trees usually become wide-spreading and rounded in full open light, they are not very compatible with long-term alley cropping because they cast too much shade on the undercrop.

Rootstocks and soils

Oaks are almost always grown as seedling trees on their own roots, as there are very few cultivars selected for good nut production.

Oaks generally prefer deep, fertile soils (often a moist medium or heavy loam) with moderate drainage and 40-150cm (16-60") rainfall. Many species tolerate a range of pH, from moderately acid to moderately alkaline. However, in this large family certain species can withstand many different conditions: poor, dry, acidic sandy soil; heavy, damp clay soil; marshy areas and dry chalk; as well as extreme aridity, salinity, alkalinity, flooding, and severe heat and cold.

Pollination

Most species take at least 15 years before they start to flower and fruit, sometimes up to 25 years, though there are exceptions: English oak (*Q. robur*) usually starts well before this, and some individuals flower at 3-5 years old.

Most oak trees are not self-fertile. Although they are monoecious (bearing both male and female flowers), the flowers ripen at slightly different times with little overlap, so always plant two or more trees to achieve pollination.

Feeding and irrigation

None usually required.

Pruning

Generally not required, apart from pruning off low branches as trees grow for access beneath.

Flowers of holm oak (*Q. ilex*).

Pests

Numerous insect pests are associated with oaks, but none cause serious damage.

Rabbits can browse on foliage and young trees should be protected with guards.

Squirrels are a serious pest, eating some acorns and burying many others. Crows will also take acorns.

Diseases

Oaks are affected by numerous diseases, but few are serious.

In North America, oak wilt caused by the fungus *Ceratocystis fagacearum* has caused heavy losses – red oaks are most susceptible, white oaks less so. The fungus is spread by oak bark beetles, and felling of infected trees is the only control.

Chestnut blight (*Cryphonectria parasitica*) can sometimes cause branch death and occasional tree death in mainland Europe and North America.

Honey fungus (*Armillaria* spp.) can occasionally cause tree deaths.

Powdery mildew is common on oak leaves in nurseries, where it can sometimes cause significant damage. On established trees it is not usually very significant.

Harvesting and yields

Harvest acorns in autumn (usually mid- to late autumn) when they fall to the ground. The first flush to fall are usually infested with weevils and should be discarded (destroy if you are worried about pest levels building up). You can lay sheets or taps on the ground to collect the acorns as they fall, or use an acorn-sized Nut Wizard to pick up nuts. Where there is little undergrowth, or in urban areas where there is concrete beneath the trees, you can just pick acorns off the ground if it is clean.

In heavy mast years (when a very large crop of acorns is produced, normally every 2-3 years for most oaks), there may be so many acorns produced that it is relatively safe to allow them to lie for weeks or months below the trees, harvesting a few at a time. If some remain there in the spring, they can still be harvested, even if they are sprouting (as long as the seed kernel hasn't turned green and the sprout is under 5cm/2" long).

Yields from individual trees can be 50-500kg (110-1,100lb) or more in a mast year, and in open oak forests up to 7t/ha (2.8t/acre).

Processing and storage

Ideally store the harvested acorns for two weeks in cool, dry conditions before using, to allow them to ripen fully and thus minimize the tannin content (see below). They also dry slightly, which makes shelling much easier.

On a small scale, the best way of shelling is to cut the acorns in half (lengthwise) with a sharp knife and use the point of the knife to prise out each half of the kernel. This method doesn't take long to prepare enough kernels to use in a meal. If the acorns are sprouting, the shell will have split and can be pulled apart, and the sprout itself should be discarded. Another method that may work with some species is to soak the acorns in water overnight, causing the shell to split open, making it easy to remove by hand. Any mouldy kernels should be discarded. At this stage, the shelled acorn kernels can be frozen, if desired, to be ground and leached at a later date. Nobody has taken up the challenge of larger-scale processing of acorns in recent times.

Most oaks produce acorns with moderate to high levels of tannins, which must be leached out before eating (see box on page 192). A few species, and occasional isolated trees of others, can produce sweet acorns with low enough tannin levels that they can be used whole in cooking, etc. (see species table on page 195).

Tannins are astringent and bitter compounds that are widely distributed in plants, where they play a role in plant protection. The astringency causes an unpleasant dry, puckering sensation in the mouth, making it impossible to eat tannic foods. In some fruits (e.g. persimmon) and nuts like acorns, the quantity of tannins decreases as ripening occurs. The tannins that cause the bitterness in most acorns are tannic acid, gallic acid and pyrogallol. The concentration of these is 1½ to 3 times higher in green (immature) acorns than in mature, ripe acorns. Tannin content varies widely (near zero to 9 per cent or more) from species to species, and also across individuals within a species. Some ranges found are:

White oak (*Q. alba*)   0.41-2.54 per cent
Red oak (*Q. rubra*)   3.72-4.47 per cent
Black oak (*Q. velutina*) 3.29-6.13 per cent

Acorns of the white oak group are often sweeter and less tannic (0.7-2.1 per cent tannins) than those of the black oak group (6.7-8.8 per cent), but species of both groups have been highly regarded as food sources, and there is considerable variation between individual trees of each species. Acorns of the black oak group are generally higher in fats than those of the white oak group.

Most acorns do not store for very long – 2-6 months. If they dry out to 60 per cent of their fresh weight, they will normally die. Acorns will keep in reasonably good condition for up to 6 months if you put them in a cool, moist, rodent- and squirrel-proof store, where the acorns can be piled in layers up to 15cm (6") thick. Turn the layers regularly to prevent mould growth.

Acorns of very hardy species (hardy to zone 3-5/6) can be deep-frozen for several years. First, they must be slightly dried to 80 per cent of their fresh weight, to prevent ice damage to the cells. Also, instead of being frozen immediately, they should be given a few days at 0°C (32°F), then a few days at -5°C (23°F), then finally -20°C (-4°F). After storage, if you want to grow them they should be cold stratified for 3-5 months before sowing.

Shelled acorn (half kernels)

Acorn meal before leaching.

Preparing acorn meal

Even when fully ripe, most acorns have too many tannins to be edible. However, they are easy to remove, taking no more time and effort than it takes to sprout seeds, and the resultant acorn meal is nutritious and flavourful.

First, use a coffee grinder or food blender to grind the acorn kernels into an acorn meal, creating small pieces (6mm/¼" or so). Then leach the acorn meal using either of the following methods:

Cold water method
Traditional methods of leaching ground acorn meal include placing them in a sack in a stream of running water, for a few days. Running water leaches tannins faster than still water, but a handy stream isn't always available, so follow these steps.

❋ Half fill a large jar with the acorn meal. Fill the jar up to the top with fresh water and put it in a cool or cold room, or in a fridge.

❋ Twice a day for 3-4 days, empty the water from the top half of the jar and refill with fresh water. Initially the water will be brown as the tannins leach out of the acorns, but it will become progressively clearer.

❋ After 3-4 days, the tannins will have gone and you can drain the acorn meal to use. The meal will need cooking before eating and can be used in many savoury or sweet dishes. Alternatively, it can be made into a flour (see opposite).

Boiling method
You can leach tannins quicker by repeatedly boiling the acorn meal and draining. However, you are likely to lose a lot more nutrients this way, and the acorn meal also ends up a much darker brown colour. The acorn meal may be cooked enough that it can be used immediately without further cooking.

Acorn meal during the leaching process – note the tannins making the water brown.

Milling oak flour.

Acorn starch

A by-product of the cold water method is acorn starch, which is used in some recipes, particularly from Korea. To save the starch, follow these steps.

❋ As you pour off the tannin water from the soaked acorn meal, pay attention to a creamy, cloudy substance in the water – this is the starch. Stop pouring once this starts to flow, top up with fresh water and soak again for a few hours. To keep the starch, you have to leach the acorns for longer (as you are not pouring off as much at each change of water), ensuring that you don't pour any of the cloudy starch out in the process.

❋ After a few days, when the acorns are ready/sweet, fill the container once more with water, stir well and allow to settle for 1 hour. Pour off the clear water and pour the cloudy water into a separate bowl. Add fresh water to the acorns and allow to settle again, adding any cloudy water from this second round to the first.

❋ Leave the starchy water to sit for 1 hour and separate. Pour off the clear water and you should be left with an off-white, slightly sticky substance, which can be dried in a very low oven or dehydrator.

Acorn flour

Acorn flour can replace a third of ordinary grain flours in almost any recipe. To make the flour, spread the acorn meal on a non-stick baking sheet and dry it in warm conditions. Then run the dried meal through a manual or electric grain mill.

Propagation

Propagation is normally by seed, which is not dormant and should be sown immediately in autumn. Be aware that the first seeds to fall are often infected with weevils. With many species, seed is only produced in quantity every few years (a 'mast' year), with very light or no crops between as an ecological defence against acorn pests and diseases.

Germination often starts in winter, with roots growing while the growing medium is still warm. Most oaks are taprooted and should be sown in deep cell containers or deep pots. Shoot growth begins in spring, but after one year the shoot-to-root ratio can be as much as 1:3. Direct sowing acorns in situ prevents transplant shock, but rodents and birds are very fond of the nuts and can decimate sowings outside.

Grow seedlings for 1-2 years before transplanting out. Smaller trees transplant much better than larger ones.

Grafting is occasionally used to propagate ornamental cultivars and could in theory be used for good fruiting selections. Side grafts are often used; evergreens are difficult.

Species

Oaks are almost always grown as seedling trees. Very little work has been done on selecting improved fruiting forms – one exception is burr oak (*Q. macrocarpa*) 'Ashworth' in the USA.

When choosing species, you will need to consider tannin levels, which vary largely with species, as well as climate and soil requirements. Recommendations are listed on pages 195-201.

Acorns of (clockwise from top left) Turkey oak (*Q. cerris*), scarlet oak (*Q. coccifera*), burr oak (*Q. macrocarpa*), English oak (*Q. robur*), sessile oak (*Q. petraea*) and holm oak (*Q. Ilex*).

Key to oak species tables

D/E: D = deciduous, E = evergreen
Ripe: indicates the number of growing seasons for acorns to ripen
Habit: LT = large tree (over 18m/60'), MT = medium tree (10-18m/33-60'), ST = small tree (under 10m/33'), T = tree, Shr = shrub, LShr = large shrub. Often can be a range, e.g. Shr/ST.
Mast: indicates length of years between heavy crops (mast years).
Length: acorn length.
Width: acorn width.

Recommended oak species with low-tannin acorns

Species	Origin	Common names	D/E	Ripe	Z/H	Habit	Mast	Length	Width
Q. agrifolia	California	California live oak	E	1	8/5	LShr		20-35mm (¾-1⅜")	10-15mm (⅜-⅝")
Q. alba	Eastern USA	White oak, stave oak, Quebec oak, American white oak	D	1	4/7	LT	4-9	10-30mm (⅜-1⅛")	
Q. arizonica	South-west USA, north-west Mexico	Arizona white oak	SE	1	7/6	Shr/ST		20-25mm (¾-1")	
Q. aucheri	South-west Turkey, Greek islands	Boz pirnal oak	E	2	8/5	LShr to 5m (16')		15-20mm (⅝-¾")	10-15mm (⅜-⅝")
Q. x bebbiana	Eastern USA	Bebb's oak	D	1	4/7	LT	1		
Q. bicolor	North-east North America	Swamp white oak, white oak	D	1	4/7	LT	3-5	20-30mm (¾-1⅛")	15-25mm (⅝-1")
Q. chrysolepis	South-west USA, north-west Mexico	Canyon live oak, canyon/maul oak	E	2	7/6	LShr/MT		25-35mm (1-1⅜")	15-30mm (⅝-1⅛")
Q. douglasii	California	Blue oak, iron oak	D	1	7/6	Shr/T		20-35mm (¾-1⅜")	10-15mm (⅜-⅝")
Q. dumosa	California	California scrub oak, scrub oak	SE	1	8/5	Shr to 4m (13')	2-3	20-30mm (¾-1⅛")	
Q. emoryi	South-west USA, Mexico	Emory oak, Western black oak	D	1	7/6	ST		15-20mm (⅝-¾")	8-10mm (¼-⅜")
Q. gambelii	South-west USA	Gambel oak, shin oak	D	1	4/7	Shr/ST to 8m (26')		15-20mm (⅝-¾")	12-15mm (½-⅝")
Q. gramuntia	Southern France	Holly-leaved Grammont oak	E	1	8/5	ST		20-50mm (¾-2")	10-15mm (⅜-⅝")
Q. ilex	Mediterranean	Holm oak, holly oak	E	1	7/6	LT		15-30mm (⅝-1⅛")	10-15mm (⅜-⅝")

Recommended oak species with low-tannin acorns

Species	Origin	Common names	D/E	Ripe	Z/H	Habit	Mast	Length	Width
Q. ilex var. ballota	North Africa, southern Spain	Ballota oak	E	1	7/6	MT		20-50mm (¾-2")	10-15mm (⅜-⅝")
Q. ithaburensis	Syria, Israel/ Palestine	Israeli oak	SE	2	7/6	ST		25-50mm (1-2")	20-30mm (¾-1⅛")
Q. ithaburensis subsp. macrolepis	South-east Europe	Vallonea oak, camata	SE	2	7/6	ST		25-50mm (1-2")	20-30mm (¾-1⅛")
Q. kelloggii	Western USA	California black oak, Kellogg oak	D	2	8/5	Shr/T, 5-25m (16-82')	2-3	25-30mm (1-1⅛")	
Q. lobata	California	Valley oak, California white oak	D	1	7/6	Shr/T	2-3	30-55mm (1⅛-2¼")	12-20mm (½-¾")
Q. lyrata x virginiana	Hybrid	Compton's oak	D	1	7/6	T			
Q. macrocarpa	Eastern USA	Burr oak, mossy cup oak, blue oak	D	1	3/7	LT	2-3	25-40mm (1-1⅝")	10-30mm (⅜-1⅛")
Q. macrocarpa x gambelii	South-west USA	Bur Gambel oak	D	1	4/7	ST	1	Large	
Q. macrocarpa x muehlenbergii x robur	Hybrid	Ooti oak	D	1	6/7	MT		Large	
Q. mongolica	Japan, Sakhalin (north Pacific Ocean)	Mongolian oak	D	1	3/7	LT		20mm (¾")	12mm (½")
Q. muehlenbergii	Eastern USA	Chinkapin oak, yellow chestnut oak	D	1	4/7	LT		13-20mm (½-¾")	12-14mm (½-⅝")
Q. prinoides	North-east & central USA	Dwarf chinkapin oak, chinkapin oak	D	1	5/7	Suckering Shr/T to 4m (13')		10-15mm (⅜-⅝")	8-10mm (¼-⅜")
Q. prinus	Eastern USA	Chestnut oak, basket oak, rock oak	D	1	5/7	LT	2-3	25-40mm (1-1⅝")	15-25mm (⅝-1")
Q. x schuettes	Eastern USA	Schuette's oak	D	1	4/7	LT		Large	
Q. stellata	Eastern USA	Post oak, iron oak	D	1	5/7	ST/MT	2-3	10-25mm (⅜-1")	6-10mm (¼-⅜")

Recommended oak species with low-tannin acorns

Species	Origin	Common names	D/E	Ripe	Z/H	Habit	Mast	Length	Width
Q. vacciniifolia	Western USA	Huckleberry oak	D	2	6/7	Prostrate/ erect Shr, 0.5-1.8m (1'8"-6')		12-15mm (½-⅝")	
Q. virginiana	South-east USA	Live oak, Virginia live oak	E	1	7/6	Shr/T	1	25-30mm (1-1⅛")	5-13mm (⅛-½")

Recommended oak species with moderate to high tannin acorns

Species	Origin	Common names	D/E	Ripe	Z/H	Habit	Mast	Length	Width
Q. acuta	Japan, North Korea, China	Japanese evergreen oak	E		7/6	LShr		15-20mm (⅝-¾")	8-12mm (¼-½")
Q. acutissima	Japan, Korea, China	Sawtooth oak, Korean oak	D	2	5/7	LT	1	20-25mm (¾-1")	
Q. afares	Algeria	Afares oak	D		5/7	MT		35-50mm (1⅜-2")	
Q. aliena	Japan, Korea	Oriental white oak	D	1	5/7	LT		20-25mm (¾-1")	
Q. alnifolia	Cyprus	Golden oak	E	2	8/5	Shr to 2m (6'6") / T to 8m (26')		25-35mm (1-1⅜")	
Q. brantii	Kurdistan, Iran		D	2	7/6	Shr/ST to 10m (33')		20-30mm (¾-1⅛")	15-20mm (⅝-¾")
Q. castaneifolia	Caucasus, Iran	Chestnut-leaved oak	D	2	6/7	LT		20-30mm (¾-1⅛")	10-15mm (⅜-⅝")
Q. cerris	Central & southern Europe	Turkey oak	D	2	6/7	LT		30-35mm (1⅛-1⅜")	
Q. coccifera	Mediterranean	Kermes oak, grain oak	E	2	6/7	Bushy Shr, 0.3-1.5m (1-5')		15-30mm (⅝-1⅛")	8-20mm (¼-¾")
Q. coccinea	Eastern USA, southern Canada	Scarlet oak, Spanish oak	D	2	4/7	LT	3-5	15-25mm (⅝-1")	15-25mm (⅝-1")
Q. dentata	Japan, Korea, China	Japanese emperor oak, daimio oak	D	1	5/7	LT		12-24mm (½-1")	12-20mm (½-¾")

Recommended oak species with moderate to high tannin acorns

Species	Origin	Common names	D/E	Ripe	Z/H	Habit	Mast	Length	Width
Q. ehrenbergii	Syria, Lebanon		D		7/6	Shr/ST		30-40mm (1⅛-1⅝")	20-30mm (¾-1⅛")
Q. ellipsoidalis	North-east USA	Northern pin oak, Jack oak	D	2	4/7	LT		12-20mm (½-¾")	10-15mm (⅜-⅝")
Q. engelmannii	California		E		8/5			20-25mm (¾-1")	
Q. faginea	North-west Africa	Portuguese oak	SE	1	7/6	Shr/ST		25mm (1")	
Q. falcata	South-east USA	Southern red oak, swamp red oak	D	2	6/7	LT	1-2	12-14mm (½-⅝")	10-15mm (⅜-⅝")
Q. frainetto	Balkans, southern Italy, Turkey	Hungarian oak, Italian oak	D	1	6/7	LT		20-35mm (¾-1⅜")	10-12mm (⅜-½")
Q. fruticosa	Southern Spain/ Portugal, Morocco		SE	1	8/5	Shr, 0.3-2m (1'-6'6"), often carpets		10-15mm (⅜-⅝")	8-10mm (¼-⅜")
Q. garryana	Western North America	Oregon white oak, Garry oak, Oregon oak	D	1	5/7	ST/MT	2-3	20-25mm (¾-1")	15-20mm (⅝-¾")
Q. glandulifera	Japan, Korea, China	Konara oak, gland-bearing oak	D	1	5/7	MT		10-15mm (⅜-⅝")	
Q. glauca	Japan, China	Blue Japanese oak	E	1	7/6	Shr/MT		10-15mm (⅜-⅝")	7-8mm (¼")
Q. haas	Asia Minor		D	1	5/7	LT		18-50mm (¾-2")	18-20mm (¾")
Q. hartwissiana	Asia Minor, Caucasus		D	1	5/7	LT		18-30mm (¾-1⅛")	10-15mm (⅜-⅝")
Q. x heterophylla	Eastern USA	Bartram's oak	D	2	5/7	LT		20-30mm (¾-1⅛")	
Q. x hispanica	Southern Europe	Lucombe oak, Spanish oak	SE	2	6/7	Shr/LT		30-40mm (1⅛-1⅝")	20mm (¾")
Q. imbricaria	Eastern & central USA	Shingle oak, laurel oak	D	2	5/7	MT/LT	2-4	10-15mm (⅜-⅝")	12-16mm (½-⅝")
Q. infectoria	Asia Minor, Middle East	Aleppo oak	SE	1	6/7	ST to 4m (13')		20-25mm (¾-1")	12-18mm (½-¾")
Q. x kewensis	Hybrid		E	2	6/7	ST		20-25mm (¾-1")	10-15mm (⅜-⅝")

Recommended oak species with moderate to high tannin acorns

Species	Origin	Common names	D/E	Ripe	Z/H	Habit	Mast	Length	Width
Q. laevis	South-east USA	American turkey oak, scrub oak	D	2	6/7	Shr/T, 6-12m (20-40')		20mm (¾")	
Q. lamellosa	Himalayas		E		8/5	Shr/T		30-40mm (1⅛-1⅝")	20-30mm (¾-1⅛")
Q. leucotrichophora	Himalayas		E	1	8/5	Shr/T		20-25mm (¾-1")	13-15mm (½-⅝")
Q. x libanerris	Hybrid	Libanerris oak	D	2	6/7	LT		25-35mm (1-1⅜")	
Q. libani	Middle East, Turkey	Lebanon oak	D	2	6/7	Shr/T to 7-8m (23-26')		25mm (1")	25mm (1")
Q. lyrata	Central & southern USA	Overcup oak, swamp post oak	D	1	5/7	LT	3-4	15-25mm (⅝-1")	15-25mm (⅝-1")
Q. macroocarpa x robur	Hybrid	Bur English oak	D	1	6/7	LT	3-5	Large	
Q. macrocarpa x turbinella	Hybrid	Burlive oak	D	1	3/7	LShr/ST		Large	
Q. marylandica	Central & south-east USA	Blackjack oak, Jack oak	D	2	5/7	ST, 6-10m (20-33')		10-20mm (⅜-¾")	
Q. michauxii	South-east USA	Swamp chestnut oak, cow oak, basket oak	D	1	6/7	LT	3-5	30mm (1⅛")	20-25mm (¾-1")
Q. myrsinaefolia	Japan, China, Laos		E	1	7/6	Shr/MT		17-25mm (¾-1")	7-10mm (¼-⅜")
Q. nigra	South-east USA	Water oak, possum oak	D	2	6/7	LT	1-2	10-15mm (⅜-⅝")	10-15mm (⅜-⅝")
Q. nuttallii	Southern USA	Nuttall's oak	D	2	6/7	LT	3-4	20-30mm (¾-1⅛")	
Q. oblongifolia	South-west USA, north-western Mexico	Mexican blue oak, Western live oak	E	1	7/6	Shr/ST to 8m (26')		15-20mm (⅝-¾")	
Q. palustris	North-east USA, south-eastern Canada	Pin oak, Spanish oak, swamp oak, swamp Spanish oak	D	2	4/7	LT	1-2	12-17mm (½-¾")	9-15mm (⅜-⅝")

Recommended oak species with moderate to high tannin acorns

Species	Origin	Common names	D/E	Ripe	Z/H	Habit	Mast	Length	Width
Q. pedunculiflora	Asia Minor, Balkans		D	1	6/7	LT		20-30mm (¾-1⅛")	15-20mm (⅝-¾")
Q. petraea	Europe	Sessile oak, durmast oak	D	1	4/7	LT	3-8	20-30mm (¾-1⅛")	15-25mm (⅝-1")
Q. phellos	South-east USA	Willow oak	D	2	6/7	LT	1	10mm (⅜")	
Q. phillyreoides	China, Japan	Ubame oak, peach oak, pin oak	E	2	7/6	Shr/ST, 3-9m (10-30')		15-20mm (⅝-¾")	8-10mm (¼-⅜")
Q. pubescens	Southern Europe	Downy oak, pubescent oak	D	1	5/7	MT/LT		15-20mm (⅝-¾")	8-12mm (¼-½")
Q. pungens	South-west USA	Sandpaper oak	E	1	7/6	Shr/T		10mm (⅜")	
Q. pyrenaica	South-west Europe, Morocco	Pyrenean oak, Spanish oak	D	1	7/6	Shr/MT		15-30mm (⅝-1⅛")	
Q. robur	Europe	English oak, pedunculate oak	D	1	6/7	LT	2-6	15-30mm (⅝-1⅛")	
Q. robur x alba	Hybrid	English white oak	D	1	3/7	LT			
Q. robur x lobata	Hybrid	Robata oak	D	1	3/7	LT			
Q. robur x turbinella	Hybrid	English live oak	D	1	3/7	Shr/ST			
Q. rubra	Eastern North America	Red oak, northern red oak, American red oak	D	2	3/7	LT	2-3	20-30mm (¾-1⅛")	15-25mm (⅝-1")
Q. semicarpifolia	Himalayas		E	2	8/5	Shr		20-30mm (¾-1⅛")	25mm (1")
Q. shumardii	South-east USA	Schumard oak, Schneck oak	D	2	5/7	LT	2-3	18-30mm (¾-1⅛")	
Q. suber	Mediterranean	Cork oak	E	1	8/5	ST/LT	2-4	20-45mm (¾-1¾")	14-18mm (⅝-¾")
Q. trojana	South-east Europe, Turkey	Macedonian oak	D	2	6/7	ST		27-45mm (1-1¾")	20mm (¾")
Q. undulata	South-west USA, Mexico	Wavyleaf oak	D	1	5/7	Shr, 1-3m (3'3"-10'), rarely T to 9m (30')		15-20mm (⅝-¾")	8-15mm (¼-⅝")

Recommended oak species with moderate to high tannin acorns

Species	Origin	Common names	D/E	Ripe	Z/H	Habit	Mast	Length	Width
Q. variabilis	China, Japan, Korea	Chinese cork oak, Oriental oak	D	2	4/7	LT	2	15-20mm (⅝-¾")	
Q. velutina	Eastern North America	Black oak, smooth-bark oak	D	2	4/7	LT	2-3	15-25mm (⅝-1")	15-20mm (⅝-¾")
Q. wislizeni	California	Interior live oak, Highland live oak	E	2	8/5	Shr	5-7	20-35mm (¾-1⅜")	10-15mm (⅜-⅝")

Where no details are given this is because information is not readily available for these species.

When considering oaks that are suitable for your location, oak is such a large family that you'll find species appropriate for every kind of climate and soil type. Here are some recommendations:

* **Cool maritime climates**: Oaks from the continental eastern USA, China and Japan don't grow or fruit so well and can suffer autumn frost damage from unripened shoots. Of these, the best are *Q. rubra* and some of the other red oaks. However, Mediterranean oaks thrive in cool temperate climates, growing faster here than in their native areas. Low-tannin species include *Q. agrifolia, Q. ilex, Q. ilex* var. *ballota, Q. ithaburensis macrolepis, Q. kelloggii*; and possibly *Q. douglasii, Q. dumosa, Q. gramuntia* (a confused species – may be part of *Q. ilex*), *Q. lobata* and *Q. vacciniifolia*. Higher-tannin species include *Q. cerris, Q. coccifera, Q. frainetto, Q. fruticosa, Q. x hispanica, Q. x kewensis, Q. libani, Q. palustris, Q. petraea, Q. phillyreoides, Q. robur, Q. rubra, Q. suber, Q. wislizeni*; and possibly *Q. alnifolia, Q. engelmannii, Q. garryana, Q. haas, Q. pubescens, Q. pyrenaica, Q. trojana*.

* **Oak species for poor soils**: *Q. ilicifolia* (bear oak), *Q. laevis, Q. x libanerris, Q. marylandica, Q. prinoides*.

* **Oak species for very alkaline soil**: *Q. cerris, Q. ellipsoidalis, Q. frainetto, Q. ilex, Q. macrocarpa, Q. macrocarpa x robur, Q. muehlenbergii, Q. robur*.

* **Oak species for very acid soil**: *Q. marylandica, Q. petraea*.

* **Oak species for wet soils**: *Q. bicolor, Q. ellipsoidalis, Q. lyrata, Q. michauxii, Q. nuttalli, Q. petraea, Q. phillyreoides, Q. robur*.

* **Oak species tolerant of saline soils**: *Q. virginiana*.

* **Drought-tolerant oak species**: *Q. alba, Q. aucheri, Q. castaneifolia, Q. chrysolepis, Q. douglasii, Q. gambelii, Q. ithaburensis, Q. leucotrichophora, Q. macrocarpa, Q. macrocarpa x turbinella, Q. marylandica, Q. prinoides, Q. pubescens, Q. pungens, Q.robur x lobata, Q. rubra, Q. suber, Q. velutina, Q. virginiana*.

* **Oak species tolerant of maritime exposure**: *Q. aucheri, Q. ilex*.

* **Precocious (early fruiting) species:** *Q. acutissima* Gobbler strain (5-8 years), *Q. cerris* (5-8 years), *Q. variabilis*.

PECAN
(Carya illinoinensis)

ZONE 6-7, H5-6

Pecan is native to the USA, from the south as far north as Illinois, but it is cultivated further north still, into southern Canada. Traditional pecan cultivars are notorious for needing a long fruit development period, with hot, sunny summer weather, and in southern USA, the Mediterranean region and similar climates they can be very productive trees – and they are an important commercial crop in these areas.

Canopy of an old pecan tree.

The traditional cultivars are not a viable crop in cooler climates. Instead, the so-called 'northern' pecan cultivars have good potential. Selected and bred in northern USA and southern Canada (Ontario), these cultivars have been chosen for their tolerance of cold winter temperatures and their ability to ripen their nuts in relatively short summer seasons.

Pecans are large deciduous trees, growing 30m (100') high or more in warm climates (though a third that size in cool climates) and they can live to a great age – 400-500 years. They have deep, furrowed, irregular brownish-grey bark and tend to form upright cylindrical crowns when grown in the open. Pecan is essentially a climax forest tree, which competes with other species for space in the forest canopy.

Leaves are aromatic, alternate and pinnate with 11-17 leaflets, each 5-17cm (2-7") long. Trees come into leaf in mid-spring and may be damaged by late spring frosts, particularly in frost pockets.

Male flowers are produced on slender, drooping catkins, which arise from lateral buds on the previous year's wood; female flowers are borne in clusters on a spike at the end of the current season's shoot. Pollen is light, dry, fluffy and pale yellow, and flowers are wind-pollinated.

Fruit are borne in spikes of 3-10. The fruit is oblong, 3-8cm (1⅛-3¼") long with a slightly 4-winged outer leathery husk (shuck). Inside each fruit is a single nut, which is smooth, light brown, thin-shelled, sweet and edible. Nuts on southern varieties can be up to 5cm (2") long but on the northern varieties they are smaller, typically 2.5-4cm (1-1⅝") long.

Pecan fruit, each containing a single nut.

Southern cultivars need hot summers to ripen wood, and in cooler regions shoot dieback is common after even mild winters. Northern cultivars don't need such high summer temperatures.

Pecans have deep taproots, which securely anchor the trees if soil conditions allow.

Uses

Pecan nuts are a good source of oleic acid, thiamin, Vitamin E, magnesium, selenium, zinc, protein and fibre (see Appendix 1). While most nuts are high in monounsaturated fats, and walnuts are high in polyunsaturated fats, pecans have a blend of both. The fatty acid content is very similar to that of olive oil.

Pecans are well-known and major commercial nuts. As well as being eaten as raw and roasted snacks, they are used in numerous baked foods (cakes, breads, biscuits, pies, etc.), dairy foods (ice creams, yogurts, cheeses), confectionery, breakfast cereals, meat and fish batters, sauces and marinades, pestos and salads.

The oil can be extracted from the kernels, and is of good quality and suitable for any culinary use.

Secondary uses of pecan

Pecan oil is used in cosmetics.

Pecan sap, obtained by tapping trees (like maples for making maple syrup), can be obtained and used to drink raw, to brew with, etc. However, fruiting trees are unlikely to be used, as it might reduce cropping.

Pecan shells are a commercial commodity. They are used in tannin manufacture, to make charcoal and as abrasives in hand soaps. Ground into a powder, they are used as a filler for plastic wood and veneer wood. The shells are also burned as a heating fuel. The shell powder (flour) of various sizes is used as a soft grit in non-slip paints, adhesives, dynamite and polishing materials.

Pecan wood isn't quite as strong as that from many other hickories, but is used similarly. Hickory wood is well known for its resilience and is excellent for tool handles (hammers, picks, axes, etc.). The wood is straight-grained, coarse-textured and heavy (similar to hickory – 820kg/m^3, 51lb/ft^3). Stiff and highly shock-resistant, it has high bending and crushing strength, and excellent steam-bending properties.

Leaves of pecans, like walnuts, contain the anti-fungal chemical juglone. Native Americans used the leaves medicinally to treat fungal diseases.

Pecans are very beautiful trees and are widely used for their ornamental value. They provide shade in summer and golden leaf colour in the autumn.

Cultivation

Because of their relatively low yields, pecans are well suited to low-input, sustainable agricultural systems, where the long-lived multifunctional trees are a valuable resource for food, fuel and high-quality timber. Although

A nine-year-old northern pecan tree.

their leaves contain juglone, which can act as a growth suppressant to other plants, problems with nearby plants have rarely been reported.

Pecans need a relatively sheltered site – limb breakages in strong winds are likely.

In warm climates, pecan trees can become large and eventually require up to 15-20m (50-70') of space between trees. In cooler climates, trees grow a similar size to almond and can be spaced much closer at 6-9m (20-30') apart. You can use large gaps between trees to grow an intercrop, for example a fruit tree. In the USA, pecan orchards are often undergrazed with cattle or sheep. Traditionally, pecan orchards were culti-vated in North America with a leguminous ground cover (often crimson clover, *Trifolium incarnatum* and/or hairy vetch, *Vicia villosa*). These provided nitrogen for cropping, and good cover for beneficial insects, which controlled pecan aphid numbers. The covers were mown short before nut harvest.

Try not to break the long taproot when plant-ing, as this can set trees back. Pecans appear to be very slow-growing for their first few years, when in fact they are growing a taproot at the expense of top growth. Tubular tree shelters may be advantageous. After a few years, annual growth of 30-45cm (1-1'6") is achieved.

Because of lack of self-fertility (see right), it is common practice to plant more than one culti-var to ensure cross-pollination between culti-vars occurs. Pecan growers usually grow 2-5 cultivars, using a mixture of protandrous and protogynous types (see 'Pollination' right). Although pollen can be carried 900m (3,000') by the wind, in cool, moist climates you should assume good pollination only for 2-3 trees' dis-tance away.

Rootstocks and soils

Cultivars are grafted on to seedling pecan rootstocks.

Pecans like a deep, fertile soil. They are adapted to rich soils in floodplains, and thus do best in a fertile loamy soil – moist but well drained. But they tolerate both light and heavy soils, and acid and alkaline conditions. They have a high demand for zinc, and deficiencies are common, depending on the soil reserves.

Pollination

Flowering is in late spring. Flowers are wind-pollinated, so good weather leads to best pollination.

Most trees are not very self-fertile, with the male and female flowers being ripe at slightly different times. The length of flowering

Pecan catkins.

depends on the season and cultivar, but typically pollen is shed from male flowers for 3-7 days, and stigmas in female flowers are receptive for 7-14 days.

Although there are variations year by year, some cultivars are protogynous (female flowers mature first), while others are protandrous (male flowers mature first). Several have no overlap between flowers, so are completely self-sterile; but generally there is some overlap of female and male flowers on the same tree. When self-pollination occurs, there's a slight decrease in nut size.

Flower initiation occurs in the spring, hence poor summer weather shouldn't have such an effect on flowering the following year as it does in nut species where flower initiation occurs the summer before (e.g. walnuts).

There is the usual drop of unpollinated nutlets in early summer. The number of fruit per cluster depends on the cultivar, and varies from 2-3 in some to 8-10 in others. After pollination, the fruit grow larger for 3 months, then the kernels fill in the remaining 2 months before harvest.

Feeding and irrigation

Young trees need no extra nutrients. If trees are producing well, they will usually need feeding with a source of nitrogen. Healthy bearing trees should produce shoots 18-40cm (7-16") long – less than this and they will produce few nuts. If growth is less, fertilize with manure, compost, etc. around the drip line of the canopy – nitrogen being the nutrient most needed. If growth is more, reduce extra feed. If a nitrogen-fixing ground cover is used, then this will probably provide all the extra nitrogen needed.

Another option is to intercrop or border the pecan trees with nitrogen-fixing trees or shrubs (see Chapter 1, tables on pages 22, 34).

Pecans are deep-rooting and do not require irrigation to grow and survive. However, as with other nuts, soil moisture is a major factor in determining the average size of nuts produced. Very dry soils in mid–late summer and early autumn will have a significant effect, hence it may be worth considering irrigation if this is likely.

Pruning

Little pruning is required. Just remove branches that are too low, overcrowded, diseased or damaged by wind or heavy crops.

Pests

Squirrels and rodents are the main pests.

Aphids can sometimes be a problem too, so grow plants (e.g. umbellifers) nearby to attract aphid predators.

Diseases

Pecan scab (*Cladosporium caryigenum*) is the most economically important disease in the USA, worst in the south-east. It is unlikely to be a problem in cooler northern regions.

Coral spot fungus (*Nectria cinnabarina*) can be a problem in cooler summer regions when the new wood doesn't probably ripen. Symptoms are pinhead-sized salmon-pink pustules on dead and young twigs and branches. If seen, infected wood should be cut out; don't leave any dead branch stubs.

Pecan is resistant to honey fungus (*Armillaria* spp.).

Harvesting and yields

Unlike most other members of the hickory family, on pecans the husks split at maturity and the nuts drop freely to the ground. Pecans ripen mid- to late autumn and yields reach around 15kg (33lb) per tree. Commercial yields rarely surpass 2,270kg/ha (2,000lb/acre or 1t/acre), and yields from northern cultivars around half this. These are in-shell yields; kernel yields are roughly half these figures.

More precocious (early fruiting) cultivars start bearing nuts after 4-5 years, while others may take 6-8 years before they start bearing. Heavy-bearing trees have a tendency to become

Pecan scab on pecans.

Pecan orchard in Texas, USA.

biennial bearers. Full cropping is reached at 20-25 years of age.

The nuts will eventually fall to the ground without help, but to aid harvest (especially in windless weather), branches can be shaken or nuts knocked down with bamboo poles. The husks often remain on trees throughout the winter.

Commercial harvesting is usually facilitated by machine shakers, which shake the tree trunks and shake off the nuts. The nuts are then swept into windrows and picked up with a mechanical harvester.

Processing and storage

In cooler climates, nuts will need drying to prevent moulds. Drying also improves the nut's appearance, aroma, flavour and texture. Once dried, storage in shell is good for two years or more in cold (fridge) temperature storage.

Propagation

Seeds are not very dormant and can be sown in autumn or spring. Some winter cold usually improves germination. Seeds can be planted in the autumn in deep containers or even in the ground. Direct ground sowing leads to strongly rooted trees that don't require transplanting; use three seeds per position, sow 7cm (3") deep and protect from rodents and squirrels (perhaps use a tree guard pushed well into the soil) – thin down to one seedling during the first season. Seeds germinate much better at high temperatures (21-30°C/70-86°F).

For their first few years, young trees form a taproot with only a few lateral feeder roots, and this taproot is usually longer and thicker than the above-ground stem. If buying or raising plants, either grow them in open-bottomed containers that air-prune the taproot (the exposed root tip dies, causing the plant to produce new braching shoots). Optionally, undercut (cut through) the taproot at 20-25cm (8-10") below ground level at least a year before transplanting. Transplant trees that are 2 or 3 years of age – older trees are likely to suffer a lot of transplant shock.

Named cultivars are propagated by grafting on to seedling pecan rootstocks. Grafting pecans is difficult. Normal methods work, for example whip-and-tongue, but the graft union needs to be kept at 27°C (80°F) for 10-14 days after grafting. This necessitates the use of a hot grafting pipe or heated box/room of some kind.

Cultivars

Where cultivars are unavailable, seedlings (particularly of the northern types) offer good potential.

Southern cultivars are the main commercial types grown worldwide in southern USA and elsewhere. They require hot summers to ripen their nuts. The list on page 210 includes the main commercial cultivars, but there are many more lesser-known and local cultivars.

The 'standard' northern cultivars have medium to large nut sizes (7-9g/nut, 110-142 nuts/kg or 50-65 nuts/lb), and shelling percentages typically in the high 50s. Some are grown as early ripening cultivars in southern

Pecans from commercial southern cultivars (left, nuts have been polished) and from northern pecan cultivars (right).

pecan-growing regions. They still require fairly warm summers.

Very early northern cultivars ripen 10 days or more before 'Colby', have small to medium nuts (6-7g/nut, 142-167 nuts/kg or 65-75 nuts/lb) and kernels that are 45-52 per cent of the nut weight. They need less summer heat than the standard northern cultivars.

Ultra-early northern varieties have small nuts (5g/nut, 196 nuts/kg or 90 nuts/lb) that mature extra early (under 140 days from bud break) on a very cold-hardy tree. Most have kernels under 50 per cent of the nut weight, moderately thick shells, and need cracking in a nutcracker. They are suitable for the most northerly regions where pecans are grown.

Key to pecan cultivar table

Protan = protandrous: male flowers mature before female flowers
Protog = protogynous: female flowers mature before male flowers

(Normally use both types. If both are ticked, this indicates likely self-fertility.)
VE = very early
UE = ultra-early

Pecan Carlson #3

Pecan Lucas

Recommended pecan cultivars

Southern cultivars	Protan*	Protog*	Comments
'Caddo'	✓		Nuts medium–large, easy extraction.
'Cape Fear'	✓		Nuts large. Tree very heavy cropper.
'Cheyenne'	✓		Nuts large. Tree very prolific cropper, very low-vigour 'dwarf' tree.
'Desirable'	✓		Nuts large. Tree a moderate regular cropper.
'Elliott'		✓	Nuts medium–large. Tree early leafing, moderate cropper.
'Forkert'		✓	Nuts large.
'Kiowa'		✓	Nuts very large.
'Melrose'		✓	Nuts medium–large.
'Moreland'		✓	Nuts large.
'Oconee'	✓		Nuts large. Good cropper.
'Schley'		✓	Nuts large.
'Shawnee'		✓	Nuts large. Productive trees.
'Stuart'		✓	Nuts large. Tree late leafing in spring.
'Sumner'		✓	Nuts very large. Tree a very heavy cropper.
'Wichita'		✓	Nuts large. Tree a heavy cropper.
Northern cultivars	**Protan***	**Protog***	**Comments***
'Colby'		✓	Nuts large, fair flavour, thick shell. Tree biennial cropper, late to drop leaves.
'Don Grotian'	✓		
'Hirschi'	✓		Nuts large, good flavour, thin-shelled, easy extraction.
'Kanza'		✓	Nuts medium size, excellent flavour. Good cropper.
'Major'	✓		Nuts medium–large, good flavour, thin-shelled. Tree a good cropper.
'Old Woman'			
'Pawnee'	✓		Nuts large, good flavour, thin-shelled, good extraction. Tree a good cropper, has strong branch angles.
'Peruque'	✓		Nuts medium–large, excellent flavour, very thin-shelled, easy to extract. Tree a good regular cropper.
'Posey'		✓	Nuts medium–large, very good flavour, easy extraction. Tree late leafing, early maturing.
'Snodgrass'		✓	

Recommended pecan cultivars

VE northern cultivars	Protan*	Protog*	Comments*
'Bryce'	✓	✓	
'Campbell NC-4'			Medium size nuts, crack well.
'Campbell NC-14'			Medium size nuts.
'Chillicothe'		✓	
'Devore'		✓	Nuts small, excellent flavour, thick-shelled. Regular cropper.
'Dumbell Lake'	✓		
'Fisher'			Nuts medium size. A good regular bearer.
'G I Hackberry'		✓	
'Gibson'	✓		Nuts small–medium size, fair flavour. Tree a good regular cropper.
'Hadu 2'		✓	
'James'	✓		Nuts medium size, good extraction. Good cropper.
'Lucas'		✓	Nuts medium size, fair flavour, medium-shelled, good extraction. Tree a good regular bearer.
'Mullahy'		✓	Nuts medium–large, excellent flavour, good extraction. Tree a good cropper.
'Norton'		✓	Nuts medium size, thick-shelled.
'Ralph Upton'		✓	
'S-24'		✓	
'Shoals West'	✓		
'Starking Hardy Giant'	✓		Nuts medium–large, good flavour, thin-shelled, good extraction. Tree a moderate cropper.
'Warsaw North'		✓	
UE northern cultivars	**Protan***	**Protog***	**Comments***
'Abbott'			
'Carlson #3'	✓	✓	Nuts small, fair flavour, thick-shelled. Tree a biennial moderate cropper.
'Deerstand'			Nuts medium size, crack well. Moderately productive tree.
'Frisbie'	✓		
'Fritz Ball'			
'Fritz Flat'			
'Green Island Beaver'	✓	✓	Nuts medium size, good flavour. Good cropper.
'Martzahn'		✓	Nuts medium size.
'Snaps Early'			Nuts small–medium, good flavour, good extraction. Tree a regular moderate cropper.

* Where no details are given, this is because information is not readily available for these cultivars.

PINES

(*Pinus* spp.)

ZONE 2-9, H3-7

Pines are a large group of evergreen trees and shrubs from North and Central America, Europe and Asia. Pine nuts (edible seeds of certain pine species) have been harvested for food for thousands of years by Mediterranean peoples, Native American tribes and indigenous peoples of Siberia and the Russian Far East. Pine nuts continue to be harvested in many northern hemisphere regions today and are marketed on domestic and international markets as a gourmet product.

Wild stone pine (*P. pinea*) in Spain.

All pine trees bear edible kernels, but only a selection are large enough to be worth bothering with for human food – about 20 species in Eurasia and North America. These range from shrubs (some of the Mexican species) to large trees with straight trunks (e.g. Siberian and Korean pines), and are adapted to a variety of conditions, from the harsh cold climate of eastern Siberia to the hot dry deserts of Nevada and Mexico. Many pines can live for 100-200 years.

The five commercially important nut-producing species are:

1) Siberian nut pine (*P. sibirica*)
2) Korean pine (*P. koraiensis*)
3) Stone pine (*P. pinea*)
4) Chilgoza pine (*P. gerardiana*)
5) Pinyon pines (*P. cembroides edulis, P. c. monophylla*, etc.)

The growing worldwide demand for pine nuts indicates there is good potential for looking at new pine nut plantations of both known and lesser-known species as well as agroforestry systems incorporating these species.

Pines perform important ecological functions, for example as soil stabilizers, reducing soil erosion. Also, many nut pines make good ornamentals and are planted both within and outside their natural range.

Uses

Pine nuts are cholesterol-free, contain 48-75 per cent fat (most of which is unsaturated), multiple micronutrients and vitamins; nutrient levels vary greatly according to species (see table on page 214). They can be eaten raw or cooked, though species with nuts high in carbo-hydrates (starch), such as singleleaf and parry pinyon (*P. cembroides monophylla/quadrifolia*), are best eaten cooked. Pinyon nut protein is easily digested and contains all 20 essential amino acids. In the USA, pinyons have been used as a food source for centuries and demand always exceeds supply.

Pine nuts are used worldwide as a nutritious healthy snack (raw or roasted) and add valuable protein to salads. If cooked, they are usually roasted but they can be included in dishes (e.g. many oriental and Mediterranean dishes) or added to confectionery, for example gourmet chocolates. The low-carbohydrate species make an excellent nut butter simply by mashing up the nuts. Another use, which is popular in Siberia, is to steep nuts in vodka to make a stimulating tonic!

Pine nut oil is obtained by pressing and is available as an expensive gourmet cooking oil. The by-products of the oil pressing are pine nut flakes, which are used in confectionery. The flakes still contain up to 30 per cent oil. When crushed further to extract oil, they turn into pine nut meal or flour, which has a range of culinary uses and is used as a livestock feed. It can be substituted for wheat or rye flour in pastries and pancakes, etc., giving them a rich nutty flavour. Mixed with water, the meal becomes a pine nut milk with a rich sweet nutty flavour.

Secondary uses of pine nut

Pine nut oil is available as a medicine (in capsules). Cold pressing in all-wood presses is preferred to retain the medicinal properties of nuts and derive high-quality oil. Although little research has been done, it is thought

Variations in pine nut nutrient levels

Species	Common name	Protein (%)	Fat (%)	Carbohydrate (%)	High vitamins
P. cembra	Arolla pine	19	59	17	
P. cembroides	Mexican pinyon	19	65		
P. cembroides edulis	Pinyon pine*	14	62-71	18	A, B
P. cembroides monophylla	Singleleaf pinyon	10	23	54	
P. cembroides quadrifolia	Parry pinyon	11	37	44	
P. gerardiana	Chilgoza pine	14	51	23	
P. koraiensis	Korean pine	17	65	12	B
P. lambertiana	Sugar pine		High	Low	
P. pinea	Stone pine*	34	48	7	
P. sabiniana	Digger pine (syn. ghost pine)	28	54	8	
P. sibirica	Siberian nut pine	19	51-75	12	B, E

* See Appendix 1 for more detailed nutritional content.

that Siberian pine nuts yield oil of the highest medicinal value. This has been used traditionally to cure a wide array of ailments, and is either ingested (decreasing blood pressure, boosting immune system resistance, etc.) or applied externally to treat dermatological disorders. Pine nut oil contains pinolenic acid, a polyunsaturated fatty acid, and is marketed in the USA as a means to stimulate cell proliferation, prevent hypertension, decrease blood lipid and blood sugar, and inhibit allergic reactions.

Pine nut oil is also used in cosmetics and as a massage oil, and has a number of speciality uses, such as a wood finish, paint base for paintings and treatment of fine skins in the leather industry.

In addition to being important nut producers, a number of pines are valuable timber species, especially Siberian nut pine (P. sibirica) and Korean pine (P. koraiensis), which are valued for their fragrance, durability, dense grain and rot resistance. They are widely used in wood carving, flooring, and expensive furniture making.

Pines are hugely important for the tapping of pine resin, which is usually distilled to give turpentine and rosin. These two products are used widely as they are, and in industry they form ingredients of many other products, including pharmaceuticals, printing inks, gums, resins, glues, asphaltic products, plastics, paints, varnishes, lacquers, disinfectants, polishes and soaps, etc.

Some pines are of exceptional cultural, symbolic and spiritual importance. Pinyon pines appear in Native American lore, while Russian monks planted Siberian pine in monasteries as sacred trees and distributed seeds to pilgrims for planting. (The importance of Siberian pine as a cultural and spiritual symbol has been popularized in Vladimir Megré's 'Ringing Cedars' book series.)

Cultivation

It is important to choose species that are well adapted to the local climate. Siberian and Korean pine, two major nut producers and timber trees of eastern Siberia, are adapted to northern regions (e.g. Scandinavia, Canada), but they do poorly in Mediterranean or temperate climates. Similarly, Mexican species are poorly adapted to temperate maritime climates. Full sun is essential for good cropping. Stone pine (*P. pinea*) is the most important European source of pine kernels for commerce, being especially valuable in Spain (Huelva), Portugal and Italy (Marches, Tuscany and Abruzzi). The seeds are called pignolias and due to their high quality, they achieve higher prices than alternatives such as Korean pine (*P. koraiensis*). They have been used for many centuries: shells have been found in Britain in the refuse heaps of Roman encampments. In Spain and Portugal, the main nut-producing countries, a total of over 550,000ha (1,375,000 acres) of stone pine exists, nearly all still harvested from forest where no cultivation techniques are applied except seeding or planting of new stands, thinnings for stand density regulation, and some pruning in the lower or inner crown to ease manual harvesting or (historically) for fuel wood.

Planting distances vary widely with species. Some remain shrubby and can be planted as close as 4m (13') apart, while large trees can require 15m (50') spacing or more in the long term. Such wide spacing gives opportunities for intercropping with other pines or trees/crops of other kinds for a number of years or decades (though trees must be removed before crowding starts).

In many regions, stone pines are the most viable species to grow for pine nuts. They are easy to grow, being free of pests and diseases and tolerating most soils. The trees thrive in wind and are often planted in shelterbelts with plants around 5m (16') apart. If they are grown specifically for pine-nut production, then plant at 6-10m (20-32') apart.

Stone pine cones are produced on the vigorous upright-growing shoots at the top of the crown. Growing stone pine in a dense stand, with trees too close together, leads to conical-shaped canopies (much like other pines when grown densely) with few vertical shoots apart from the leader, and thus few cones. Only vigorously growing, widely spaced stone pine trees with the characteristic umbrella-shaped canopy will render high yields. Trees are slow-growing.

Stone pine prefers loose, sandy soils, though it produces well on stony sites or gravels. Compact layers of clay or silt will restrict roots and crown development. Flowering is late, normally avoiding frost damage. Annual rainfall of over 600mm (2') is best for fully sized cones and nuts. Cones are produced from about 10 years onwards, with full crops produced by 40 years.

Flowers on Korean pine (*P. koraiensis*).

A heavy crop (mast) is produced every 3-4 years. Each cone holds about 50-100 nuts and each cone contains 20 per cent of nuts by weight.

Rootstocks and soils

Pines are almost always grown as seedling trees on their own roots. Stone pine (*P. pinea*) has had some improved cultivars selected, which are grafted on to seedling rootstocks.

Pines need a well-drained soil; sandy soils are ideal, chalky soils are tolerated. They will tolerate a wide range of soil pH, except very alkaline conditions.

Pollination

Flowers are wind-pollinated, and pines are not very self-fertile, so a minimum of two trees of a particular species should be planted.

Feeding and irrigation

Neither is often required. Pines are quite drought-tolerant.

Pruning

Little is needed. Most pines in open situations grow strong branches. You will probably need to remove some low branches in the early years to gain access beneath trees.

Pests

All the usual culprits – squirrels, birds such as crows, mice and other rodents – can predate on the pine nuts. There are also a large number of insect pests that occur in different parts of the world. In the late 1990s the western conifer seed bug (*Leptoglossus occidentalis*) was accidentally imported with timber from western USA to northern Italy, and has since spread across Europe as an invasive pest species. It feeds on the sap of developing conifer cones throughout its life, and its sap-sucking causes the developing seeds to wither and misdevelop. It is now threatening stone pine (*P. pinea*) in Italy.

Diseases

Many different rust fungi can affect pines, but the two most important are:

1) **White pine needle rust** (*Cronartium ribicola*): causes a devastating disease of American white pines, including sugar pine (*P. lambertiana*). The alternate hosts are cur-

Pine attacked by white pine blister rust.

rants (*Ribes* spp.), especially blackcurrant (*R. nigrum*), which are banned from planting in some areas of North America.

2) **Red band needle blight** (syn. Dothistroma needle blight, *Dothistroma septosporum*): can cause serious defoliation and even death of trees in some regions.

Harvesting and yields

Stone pines (*P. pinea*) start cropping at 9-10 years of age. A drawback of some other pine species is that they only start bearing crops at 20-40 years of age, though they may then bear economic crops for several hundred years, albeit at irregular intervals of 2-5 years.

Most pines flower in midsummer and cones ripen in autumn the following year (i.e. second year) or the year after that (third year). In some species, cones naturally open and drop their seeds (e.g. *P. edulis*, *P. koraiensis*, *P. pinea*); in others, cones fall from the tree intact with the seeds (e.g. *P. albicaulis*, *P. cembra*). With the latter kind, cones can simply be gathered from the ground if predation isn't too bad. With the former, seeds can be collected on sheets beneath trees (shaking trees if necessary) or cones must be harvested from trees before they open. There is usually a period of about a month between cones ripening and opening, and in this period cones should be cut or knocked off the tree with a long pole/hook and collected. Some pines are adapted to release their seeds after a forest fire (e.g. *P. gerardiana*, *P. sabiniana*) and these may need to have their cones heated to open. Commercial harvesters always harvest cones from the trees, which may necessitate someone climbing the tree.

In Mexico, small trees are sometimes shaken over plastic sheets to encourage nuts to fall. Nuts are then collected by hand. (A traditional harvest method in the region is to allow kangaroo rats, of which there are large populations, to collect the nuts and store them in tunnels a few inches underground; at the end of the harvest season, these stores are raided!)

Yields from stone pine (*P. pinea*) are about 10kg/tree (22lb/tree) for mature trees, and in commercial orchards average up to 1t/ha (880lb/acre). These are in-shell yields.

Yields from stands of wild trees are less: pinyon pines (*P. edulis*, etc.) can yield up to 335kg/ha (295lb/acre), Korean pine (*P. koraiensis*) up to 100kg/ha (88lb/acre), Siberian pine (*P. sibirica*) up to 50-60kg/ha (44-53lb/acre).

A cone on stone pine (*Pinus pinea*).

Processing and storage

Most pine cones open in a dry, warm situation but this can take days or weeks. Once open, the pine nuts shake out easily. The majority of Mediterranean commercial growers send their closed stone pine cones to processing plants, where they are 'crunched' in a machine and broken into bits, before being sorted to remove the cone species from the nuts. The nuts still need to be shelled, though, and as pine nuts are smaller than most nuts covered in this book, you certainly wouldn't want to spend half your life manually shelling them! Happily nut-cracking machines, both manual and electric, shell them perfectly well (see Resources).

Unshelled nuts have excellent keeping qualities, storing for 3 years even in warm conditions, but shelled nuts should be used within 3 months. Some drying is necessary in cooler or humid climates, and after drying, pine nuts will store for several years.

The total world production of pine nuts is around 20,000 tonnes (22,000 tons) of kernels annually, though this fluctuates widely from year to year because most good harvests occur biennially. Demand is significantly greater than supply, and the world market is often completely out of stock for months before the new harvest. The consequence is a high price (pine nuts are one of the most expensive nuts on the market). The limited supply of pine nuts also means that the market can absorb fluctuations in production with very little price alteration.

Some of the pine nuts exported from China are originally harvested in Russia and brought to China for processing and packaging. Russia's vast Siberian and Korean pine forests are under heavy logging pressure, with an estimated 20 million cubic metres (700 million cubic feet) of timber harvested illegally annually, most of it shipped to China and Japan. Russian domestic demand has also increased since Vladimir Megré's 'Ringing Cedars' book series.

Propagation

Pine trees are usually grown from seed, though if and when named cultivars become available (e.g. of stone pine, *P. pinea*), grafting will be used for them: rootstocks (*P. pinea*, also *P. halepensis*, *P. pinaster*, *P. radiata*, *P. sabiniana*) need to be at least 18 months old.

Pine seeds require varying amounts of cold treatment (stratification) before they are sown – see the table opposite for stratification requirements. Seeds should be sown in a well-drained compost, preferably in deep cells or pots, covered with 1cm (3⁄8") of compost and

Pine nut size and stratification requirements

Species	Common name	Length	Width	Dormancy
P. albicaulis	Whitebark pine	10mm (⅜")	8mm (¼")	4 weeks
P. armandii	Chinese white pine	12mm (½")	10mm (⅜")	13 weeks
P. ayacahuite	Mexican white pine	9mm (⅜")	7mm (¼")	Not dormant
P. cembra	Arolla pine	12mm (½")	7mm (¼")	26 weeks
P. cembroides	Mexican pinyon	20mm (¾")	7mm (¼")	13 weeks
P. cembroides edulis (P. edulis)	Pinyon pine	20mm (¾")	7mm (¼")	4 weeks
P. cembroides monophylla (P. monophylla)	Singleleaf pinyon	20mm (¾")	7mm (¼")	4 weeks
P. cembroides quadrifolia (P. quadrifolia)	Parry pinyon	13mm (½")		4 weeks
P. coulteri	Coulter's pine	20mm (¾")	10mm (⅜")	8 weeks
P. flexilis	Limber pine	12mm (½")	9mm (⅜")	4 weeks
P. gerardiana	Chilgoza pine	22mm (¾")	8mm (¼")	13 weeks
P. jeffreyi	Jeffrey's pine	12mm (½")	8mm (¼")	8 weeks
P. koraiensis	Korean pine	16mm (⅝")	10mm (⅜")	13 weeks
P. lambertiana	Sugar pine	15mm (⅝")	10mm (⅜")	26 weeks
P. pinea	Stone pine	18mm (¾")	10mm (⅜")	4 weeks
P. sabiniana	Digger pine (syn. ghost pine)	25mm (1")		4 weeks
P. sibirica	Siberian nut pine	13mm (½")	8mm (¼")	13 weeks
P. torreyana	Torrey pine	22mm (¾")	10mm (⅜")	4 weeks

kept at about 19°C (66°F). Very high temperatures can inhibit germination. When germination occurs, a long taproot will grow before the shoot even emerges; if seedlings are to be transplanted out of a seed tray, then care must be taken not to damage the roots.

Because of the high risks of rodent damage, it isn't recommended to sow seeds outside in beds unless you are sure that rodent control is adequate. Other pests that will eat seeds if they can get to them include squirrels, birds and (in North America) chipmunks.

Cells or pots containing seedlings must have a thin layer of pine needles or soil added from beneath an established pine tree, to allow mycorrhizal infection around the roots. These symbiotic fungi are essential for plants to grow and remain healthy (see Chapter 1, page 27). Without such infection, pine seedlings will die after a couple of years.

Seedlings don't need shading, except in very hot and sunny locations. If seedlings are planted outside in nursery rows, mulch them in the autumn to avoid problems of frost heave, which can be very damaging.

Species

There are many pine species. Some are well suited to cold winter climates, others to Mediterranean-type climates, and still others are at home in mild temperate climates, as follows.

1) **Mediterranean:** the Mexican and Californian species and chilogoza pine (*P. gerardiana*).
2) **Continental:** (cold winters, hot summers): most other North American species, Korean and Siberian pines.
3) **Temperate:** arolla pine (*P. cembra*), Chinese white pine (*P. armandii*), stone pine (*P. pinea*) and whitebark pine (*P. albicaulis*).

Pines have very high genetic variability, and within a stand there are usually individuals that are prolific seeders and produce heavy crops, sometimes even annually – see table opposite. Thus there is good potential to select good nut producers and propagate them vegetatively, for example by grafting. Some selection for improved cultivars has been achieved with stone pine (*P. pinea*) in Spain but cultivars are not yet commercially available.

Pine nuts of Arolla pine (*P. cembra*), Korean pine (*P. koraiensis*), stone pine (*P. pinea*) and Siberian pine (*P. sibirica*).

Key to pine species table

Z/H: indicates hardiness zone – USDA scale / RHS scale.

Habit: LT = large tree (over 18m/60'), MT = medium tree (10-18m/33-60'), ST = small tree (under 10m/33'), LShr = large shrub. Often can be a range, e.g. ST/MT.

Cn: indicates number of growing seasons that cones (and nuts) take to ripen.

Recommended pine species for pine nuts

Latin name	Common names	Origin	Z/H	Habit	Cn*	Comments
P. albicaulis	Whitebark pine	Western North America (California to British Columbia)	3/7	ST/MT		Bark of young trees is smooth, white and peeling; branches spreading. Needles are in fives, persisting for 4-8 years, about 5cm (2") long, stiff but flexible, dark green. Cones are borne singly at the end of branches; they are oval, 4-7cm (1⅝-3") long by 4-6cm (1⅝-2½") wide, dull purple at first but ripening brown, with short, thick scales. The cones do not open when ripe, but instead fall intact from the tree; they must be broken up to release the seeds. Seeds are wingless.
P. armandii	Chinese white pine, Armand's pine, David's pine	Mountains of western and central China	5/7	LT	2	Has widely spreading, horizontal branches. Bark is thin, greyish-green. Needles in fives, persisting for 2-3 years, thin and limp, yellowish-green to bright green. Cones are borne in groups of 1-3 on stalks; they are 10-20cm (4-8") long by 4-6cm (1⅝-2½") wide, erect but becoming pendulous, cylindrical, green at first ripening yellowish-brown, with thick, woody seed scales. Seeds are reddish-brown, wingless. Seeds are regularly collected and sold in markets in its native region, and regarded as a delicacy. Flowering begins quite early, around 12 years of age. **Note: There have been reports that a small number of people may have an allergic reaction to the seeds of this species, resulting in a metallic taste in the mouth that can last for days or weeks.**
P. ayacahuite	Mexican white pine	Mexico	7/5	ST/LT		A tree often smaller in cultivation, with a spreading head of branches. Bark is light grey, smooth on young trees, becoming rough and scaly later. Needles in fives, persisting for three years, are thin and limp, glaucous green. Cones are borne singly or in clusters of 2-3 on stalks at the end of branches; they are cylindrical, curved, 25-45cm (10-18") long by 6-14cm (2½-5½") wide at the base. Cones gape open when ripe, releasing seeds which are brown with dark stripes and wings 25mm (1") long. Flowering begins at an early age. Two naturally occurring varieties, var. *brachyptera* and var. *veitchii*, have larger seeds than the type (both about 12mm/½" long).

Recommended pine species for pine nuts

Latin name	Common names	Origin	Z/H	Habit	Cn*	Comments
P. cembra	Arolla pine, Swiss stone pine	European Alps	4/7	ST/MT	3	Usually a small- or medium-size tree, growing 10-20m (33-66') high, occasionally more. It has a wide, picturesque often broken crown in the mountains but in cultivation usually narrowly conical with branches down to the ground. Bark is grey-green and smooth at first, becoming grey-brown and fissured with age. Needles in fives, in dense brush-like clusters, persisting for 3-5 years, stiff and straight, dark green. Cones are borne on short stalks at the end of branches; they are 6-8cm (2½-3¼") long by 5cm (2") wide, egg-shaped, violet becoming brown when ripe. Cones do not open, but fall from the tree with their seeds in the spring of their third year. Seeds are reddish-brown, unwinged. The Arolla pine is slow-growing to begin with and long-lived. Fruiting is usually regular by the age of 25-30 years. Seeds are not freed until the cone disintegrates, thus the cones must be mechanically broken up. Does well in open areas, on north slopes with light, well-drained soils. Susceptible to honey fungus.
P. cembroides	Mexican pinyon, Mexican stone pine, Mexican nut pine	South-western USA and Mexico	7/5	LShr/ ST	2-3	A shrub or small tree with a rounded crown, short stem and outspread branches. Needles in groups of 1-5, persisting for 3-5 years, sickle-shaped, dark green. Cones are short-stalked, round, 3-5cm (1⅛-2") long by 3-4cm (1⅛-1⅝") wide, yellowish to reddish-brown, with only a few scales that open widely when ripe. Seeds are blackish and wingless, thick-shelled. A pinyon pine, this and its subspecies are of commercial importance. They are intolerant of competition, slow-growing and very long-lived, not reaching full maturity until 250-350 years. Other minor and similar pinyon pines from the same region (some of which may also be subspecies of this) include *P. johannis* (Johann's pinyon), *P. maximartinezii* (big-cone pinyon), *P. nelsonii* (Nelson's pinyon), *P. pinceana* (Pince's pinyon) and *P. remota* (paper shell pinyon).

Recommended pine species for pine nuts

Latin name	Common names	Origin	Z/H	Habit	Cn*	Comments
P. cembroides edulis (syn. P. edulis)	Pinyon pine, piñon pine, Colorado piñon, nut pine, two-leaved nut pine, two-needle pinyon, silver pine, Rocky Mountain nut pine	South-western USA and Mexico	5/5	ST/MT	2-3	Usually multistemmed with an irregular habit. Needles in twos or threes, stiff and dark green. Cones usually open in mid- to late autumn after a frost, the seeds falling out over a period of two weeks. Seeds are thick-shelled. A slow-growing straggling tree, adapted to a dry climate (requiring 30-45cm/1'-1'6" of annual rainfall). Trees under 25cm (10") diameter appear to be dioecious (male and female flowers are borne on different plans), producing fewer cones but many seeds per cone; while larger trees seem to be monoecious (bearing both male and female flowers) and produce many cones, with fewer seeds per cone. Young trees start to bear nuts when they are about 25 years old and 1.5-3m (5-10') high; heavy crops are not borne until trees are about 75 years old. This time factor is responsible for there being no cultivated orchards of pinyons. Cone production varies greatly from year to year, with large crops every 4-7 years; cones take 2-3 years to mature and successive good seasons are required for a sizeable yield. Seeds are called pinyon nuts when sold.
P. cembroides monophylla (syn. P. monophylla)	Singleleaf pinyon	South-western USA	6-7/5-6	ST/MT	2-3	Usually multistemmed, with a flat crown. Needles are borne in ones, are thick, stiff and prickly, grey-green and striped. Cones are 5-8cm (2-3¼") long and very wide, with woody scales; they fall in winter or early spring. Another pinyon pine that has been used for food for at least 7,500 years. Drought-tolerant, it needs a hot, dry position. Reaches maturity in 100-225 years.
P. cembroides quadrifolia (syns P. cembroides parrayana, P. parrayana, P. quadrifolia)	Parry pinyon, four-leaved nut pine	California, USA	7/5	ST/MT	2-3	A pyramidal tree, becoming flat-crowned with age, it grows up to 15m (50') high, with thick, spreading branches. Needles are borne densely, usually in fours; they are short, stiff and bluish-green. Cones are nearly round, 5cm (2") across, ripening in their second year and bearing a few dark brown seeds with thin, brittle shells. Another pinyon pine, which likes an acid soil; very drought-resistant. Large seed crops occur at intervals of 1-5 years.

223

Recommended pine species for pine nuts

Latin name	Common names	Origin	Z/H	Habit	Cn*	Comments
P. coulteri	Coulter's pine, big-cone pine	California and Mexico	7/5	LT		A straight-stemmed tree with a loose, open, pyramidal crown and very stout, wide-spreading branches. The bark is thick and very dark brown. Needles are in threes, persisting for 2-3 years, long, stiff and dark bluish-green. Cones are borne on short stalks, are very large and heavy, 25-35cm (10-14") long and up to 15cm (6") wide, and a shining yellow-brown; they are very persistent. Most cones do not open to release their seeds. Seeds are black with a 25mm (1") wing. Trees are quite fast-growing and drought-tolerant.
P. flexilis	Limber pine	Western USA	2-3/7	ST/LT		A variable tree with a trunk up to 1-1.5m (3-5') in diameter. The crowns of young trees are conical, becoming broadly rounded with age; the bark is dark grey and furrowed on old trees. Needles are in fives, persisting for 5-6 years, stiff and densely crowded at the branch tips, bluish-green. Cones are borne at the end of branches, becoming pendulous, cylindrical, 7-15cm (3-6") long and 4-6cm (1⅝-2½") wide, glossy and light brown with thick, woody seed scales. Cones mature in late summer to early autumn and scales open when ripe, within about a month; seeds are reddish-brown with a rudimentary wing. A gnarled, slow-growing pine, which is wind- and drought-hardy.
P. gerardiana	Chilgoza pine, Gerard's pine, Nepal nut pine	Himalayas, Tibet, Kashmir and northern Afghanistan	7/5	ST	2	Dense, rounded crown and short, spreading branches. The bark is thin, silvery-grey and peeling. Needles in threes, green, persisting for three years. Cones are oblong, 12-20cm (5-8") long and 7-11cm (3-4½") wide, very resinous, ripening in their second year. Some 15-20 cones are produced per tree, each cone containing 100 or more seeds. After collecting, cones are traditionally heated by a fire to loosen the scales and release the nuts. The shell around the kernel is papery and much thinner than in stone pine (P. pinea).
P. jeffreyi	Jeffrey's pine	Western North America	5/7	LT		A large tree with a trunk of over 1m (3') diameter. Cinnamon-brown bark splits into plates on older trees. The crown is rounded and branches are stout, outspreading, often somewhat pendulous. Needles are in threes, clustered at the shoot tips, persisting for two years, long, stiff and grey-green. Cones are conical and short-stalked at the end of branches, 14-20cm (6-8") long and 4-8cm (1⅝-3¼") wide, spreading horizontally, light brown, with thorny seed scales. Seeds have 3cm (1⅛") wings.

Recommended pine species for pine nuts

Latin name	Common names	Origin	Z/H	Habit	Cn*	Comments
P. koraiensis	Korean pine, Korean white pine, Korean nut pine, Chinese nut pine	Manchuria, Korea and northern Japan	3/7	MT/LT	2	A pyramidal tree of loose conical habit, reaching 20-30m (65-100') high with a trunk up to 2.5m (6') in diameter. Branches are strongly horizontal to erect. Needles are in fives, loosely arranged, stiff, green one side and bluish-white the other. Cones are borne at or near the end of branches in groups of 1-3, are cylindrical and erect, 9-14cm (3½-5½") long and 5-6cm (2") wide, bright yellowish-brown when ripe, with woody scales. Cones ripen in their second year in early autumn and seeds fall a month after ripening. Seeds are greyish-brown, unwinged. Trees start to bear cones at 20-25 years of age, heavy seed years occurring every 2-3 years. Cones contain, on average, 160 seeds. Highly valued in Asia, where improved selections exist. In North America, two improved selections, 'Grimo' and 'Morgan', have been made.
P. lambertiana	Sugar pine, Lambert pine	Western North America	7/5	MT/LT	3	Very straight-stemmed, often with the first branches way up the stem. Branches are horizontal or somewhat nodding. Needles are in fives, very stiff and wide, persisting for 2-3 years, dark green. Cones are borne at the end of branches on stalks 9cm (3½") long; they are cylindrical, pendulous, 30-50cm (1'-1'8") long and 8-11cm (3¼-4½") wide (the largest of all pines), a shining light brown with leathery scales. The cones ripen and fall off in their third year. Seeds are nearly black with brown wings 2cm (¾") long. Susceptible to white pine blister rust.
P. pinea	Stone pine, umbrella pine, Italian stone pine	Mediterranean region	7/5	MT/LT		Tree with a broadly arched, umbrella-shaped crown and horizontal branches. Needles are in twos, persisting for two years, stiff and light green. Cones are usually borne singly (occasionally in twos or threes) at the ends of branches on stout stalks, inclined downwards, nearly round, 8-15cm (3¼-6") long and up to 10cm (4") wide; they ripen in their third year. Seeds are thick-shelled, dull brown with variably sized wings (3-20mm/⅛-¾"). Thin-shelled selections exist. The cultivar 'Fragalis' has thin-shelled seeds and is cultivated for this reason. Young plants are susceptible to frost damage.

Recommended pine species for pine nuts

Latin name	Common names	Origin	Z/H	Habit	Cn*	Comments
P. sabiniana	Digger pine, ghost pine	California, USA	7/5	MT/LT		An often multistemmed tree with an open, rounded crown and grey-brown, thick, deeply fissured bark. Branches are irregularly arranged. Needles are in threes, persisting for three years, long, light blue-green. Cones are oval, borne in ones, twos or threes on stalks at the end of branches; they are 15-25cm (6-10") long and 10-18cm (4-7") across, weigh up to 1.8kg (4lb) each and are reddish-brown with large woody scales armed with spines; they persist on the tree for up to seven years after the seeds have been shed. Seeds are dark brown with wings 1cm (⅜") long. A drought-resistant species. After collecting, cones are heated by a fire to loosen the scales and release the nuts. Long used for food by Native Americans of northern California. Note: The name 'digger pine' is becoming frowned upon in the USA, even though it is widely used internationally, due to its association with the derogatory term 'Digger Indians'.
P. sibirica (syn. P. cembra sibirica)	Siberian nut pine, Siberian pine, Siberian cedar, cedar pine, Russian cedar	North-eastern European Russia, Siberia and northern Mongolia	3/7	LT		A large narrowly conical tree up to 2m (6'6") in diameter, often with a more rounded crown as a young tree. Cones are 6-12cm (2½-5") long by 5-7cm (2-3") wide. Seeds are dark brown with an easily broken shell. Trees are long-lived, up to 900 years old. The tree forms a shallow root system on wet sites and a deep taproot on dry sandy and stony sites. Trees start to bear cones at 20-25 years of age. Heavy seed years occur every 2-3 years. Cones contain about 150 seeds. Called 'cedar nuts', their proteins are more easily digested than those of walnut and hazel. The oil pressed from the nuts is high in polyunsaturated fatty acids.
P. torreyana	Torrey pine, Soledad pine	Southern California	7/5	ST/MT	3	A broad, open, irregular tree similar to digger pine (syn. ghost pine – P. sabiniana). Its trunk is 30-60cm (1-2') in diameter with deeply and irregularly furrowed bark. Needles in fives, persisting for several years, are clustered at the branch tips; they are very tough, large and dark green. Oval cones are 10-15cm (4-6") long and 9cm (3½") wide on stalks, horizontal or drooping, dark violet when young and becoming a shining chocolate-brown when ripe. Cones ripen in their third year and fall off the year after. Seeds are dark brown, speckled, very large, with a ring-like wing.

Where no details are given, this is because information is not readily available for these species.

SWEET CHESTNUTS
(*Castanea* spp.)
ZONE 4-6, H6-7

The chestnut (*Castanea*) family contains species that vary from dwarf shrubs to large trees, all of which have edible nuts. The shrubby chinkapins are described earlier (see page 143), but the important economic species are the 'sweet chestnuts', comprising three tree species and their hybrids:

Sweet chestnut 'Marki' in autumn bearing a heavy crop.

* European sweet chestnut (*C. sativa*). Originating in southern Europe, Asia Minor and north Africa, European sweet chestnut is a broad-crowned tree growing to 30m (100') high or more. Hardy to Z5-6/H6-7, it can live for 250-500 years or longer. It has been in cultivation for many centuries, particularly in southern Europe for its nuts and elsewhere for its wood.

* Euro-Japanese hybrid chestnuts (*C. sativa* x *C. crenata*). Of intermediate character, these cultivars have been selected for their excellent nut quality.

* Japanese chestnut (*C. crenata*). Resembling *C. sativa*, this is a large shrub or small tree up to 9m (30') high. Originating in Japan, where the nuts have long been of economic importance. Japanese chestnut has been used in breeding programmes in Europe and North America because of its resistance to chestnut blight. Hardy to Z6/H6, it can live for several hundred years.

* Chinese chestnut (*C. mollissima*). From central and northern China, the Chinese chestnut is a medium / large-size tree growing 12-20m (40-70') high, though sometimes shrubby. Hardy to Z5/H7, it can live for several hundred years. It is of great economic importance in China, with many nuts now exported.

Chestnuts have been a major world nut crop for many years, and remain so today. The annual world chestnut production is some 500,000 tons, with the largest producers being China, Korea, Italy, Turkey, France, Spain, Portugal and Greece. All these countries have large, healthy and expanding chestnut industries. In Europe, after a low during the 1970s, increasing acreages of chestnuts have been planted, with new varieties and less rugged planting terrain reducing the role of the traditional mountain orchards.

Chestnut trees are deciduous and have alternate, parallel-ribbed, conspicuously toothed leaves, always oblong or oval in shape. All species are monoecious in flowering habit – i.e. both male and female flowers are borne on plants, but most plants are fairly self-sterile, usually because the male and female flowering periods don't overlap (the male flowers are usually earlier). Hence more than one selection is usually needed for good crops of nuts. Flowering (yellow catkins, borne from the leaf axils of young shoots) occurs around midsummer or just after, with nuts ripening within prickly burrs in mid- and late autumn. Although largely wind-pollinated, the flowers attract bees, which feed on the nectar.

Uses

Flavourful and rich in energy, sweet chestnut has a carbohydrate content that is similar to that of rice or wheat, which is why it is sometimes called a 'tree grain' (see Appendix 1 for a detailed breakdown of its nutritional contents). Prior to the introduction of potatoes, sweet chestnuts were once the basic food of the rural poor in much of southern Europe, largely replacing cereals in mountain areas unsuited to tillage (see box opposite).

Sweet chestnuts of all species can be eaten fresh, though of course some have a better flavour than others. Most chestnuts are eaten cooked, by either quickly roasting or boiling for 5-10 minutes; steamed blanched chestnuts are a favourite French method.

Sweet chestnut – 'tree of bread'

For centuries, sweet chestnut was a staple food for generations of mountain people in France and Italy; it also constituted the food of rural populations who turned to it in times of famine and poverty. Its wood was used to heat country dwellings; it provided tannin, litter and leaves for livestock as well as raw materials for buildings, pole production and items of daily use.

As sweet chestnut provided such a basic important food, it was known as the 'tree of bread' and cultivation extended beyond its natural cultivation area, where it bore fruit due to careful tending of trees. As the cultivation gradually extended, it provided an alternative to cereals as a food for the masses, due to it being easily available and easy to store. Later, thanks to its low cost and high nutritional content, it became known as the 'bread of the poor', providing energy and protein to impoverished communities.

In past times in many mountainous areas, the average diet was based on chestnuts for at least 4-6 months of the year, with consumption about 150kg (330lb) per person per year. In a self-sufficient economy, chestnut growers often planted different varieties to meet various requirements (for drying, flour making and fresh consumption). There was great creativity in inventing various ways of preparing the chestnuts: roasted or boiled in milk water, and eaten as a bread substitute; served hot with wine or milk in the form of a soup; ground and used as a cereal flour substitute for making polenta, porridge, flatbreads, chestnut breads and thick soups.

Some varieties are especially noted for making chestnut flour, and are still used for this, particularly in Corsica and Italy. Unlike most nuts, chestnuts are low in fats and hence are more akin to cereals than other edible nuts. To make flour, the nuts must be dried, then shelled and ground. The flour is often mixed with wheat flour or other foodstuffs; it is used to make a thick soup, porridge, bread, thin cakes/biscuits, chestnut fritters and pancakes, and in stews. Dried chestnuts and chestnut flour are becoming increasingly popular for soups and polenta, and are being used to prepare tagliatelli, gnocchi and ravioli.

Chestnuts have a high sugar content, so have long been used to make desserts such as *marrons glacés* (sweet glazed chestnuts), mousse, soufflé, speciality pastries and ice creams. These days the trend is for less elaborate desserts such as chestnut flour bread, fritters and milk-based puddings.

Secondary uses of chestnuts

Chestnut timber is valued for its colour, durability and ease of working. Nowadays the main wood products in the Mediterranean are related to viticulture (posts) as well as furniture and flooring. Chestnut has good eco-credentials because of its natural durability, and as dangerous preservative chemicals are gradually phased out the demand for chestnut can only increase.

By midsummer the dense sweet chestnut canopy in an orchard starts to suppress most plants underneath.

Cultivation

All species are drought-tolerant, and prefer full sun; several tolerate part shade, especially in areas with hot summers. Harvest is easier on gentle slopes, though there is a long tradition of growing chestnut on mountain slopes.

Most species are best adapted to continental climates, with hot summers and cold winters. In cooler temperate climates, European sweet chestnut (*C. sativa*) and its hybrids, along with Japanese chestnut (*C. crenata*), are the best species to grow. The hybrids are precocious (early fruiting) trees with good-sized, quality nuts that are well suited to fresh use or processing; they prefer lower altitudes and deeper soil, and are more susceptible to late spring frost damage.

Plant at a spacing of 10-12m (33-40') apart. There is scope for double-density planting or interplanting with other tree crops, thinning out after 10 years or so. Ensure there is a pollinator on a regular basis – closer in rainy climates (within three trees) than in dry climates.

Rootstocks and soils

Several of the Euro-Japanese hybrids selected in France are increasingly used as rootstocks as well as being good fruiting cultivars themselves. They are easier to propagate by stooling or softwood cuttings (see Chapter 2, page 59), are tolerant or resistant to ink disease (see page 234) and have good genetic compatibility with the best cultivars. Cultivars sometimes used for rootstocks are:

* 'Ferosacre' – resistant to ink disease
* 'Maraval' – low vigour, resistant to ink disease
* 'Marigoule' and 'Marsol' – vigorous, moderate resistance to ink disease
* 'Marlhac' – moderate vigour, moderate resistance to ink disease

All species have similar requirements: they prefer well-drained loamy soils (tolerating light, medium, heavy, poor and dry soils, but not heavy clay) and an acid or neutral pH – ideally pH 5-6 but tolerating more acid than that. They are averse to alkaline soils but sometimes tolerate some limestone soils. All prefer a high organic matter level (minimum of 2 per cent) in the soil.

Pollination

Flowers are borne from the leaf axils near the tips of shoots. The male flowers are yellow catkins; the female flowers are borne in groups of up to five at the base of bisexual catkins. The period of maximum pollen dispersion almost always occurs about a week before the period of maximum female receptivity.

Individual varieties and trees are divided into one of four flowering types, depending on the male flowers (catkins) – basically, the 'bushier' the catkin, the more fertile the pollen is:

* **A-stamen:** without stamens, hence cannot produce any pollen. Trees sterile.
* **Brachy-stamen:** stamen threads very small, 1-3mm (⅛") long; little pollen. Trees practically sterile.
* **Meso-stamen:** stamen threads 3-5mm (⅛") long; little pollen. Trees practically sterile.
* **Long-stamen:** stamen threads 5-7mm (¼") long; abundant pollen (usually fertile).

Sweet chestnut flowers. 1. In full flower, with male catkins and female flowers fully ripe. 2. At the end of flowering, catkins turn brown and the female flowers start to grow. 3. Two weeks later, the old catkins have fallen off and burrs are growing fast.

Sweet chestnut 'Vignols' in full flower just after midsummer.

Self-pollination is only possible with long-stamen-type trees, but even these pollinate better with cross-pollination.

Flowering is in summer and this means chestnut is quite reliable in terms of setting a crop every year. Pollination is via wind as well as bees and other insects (butterflies, beetles and syrphids) – beehives are often moved into chestnut orchards in some regions, with the resulting chestnut honey a high-value speciality. Cold or wet weather throughout the flowering period can lead to very poor pollination and subsequent nut production. Warmth is essential during flowering for fertilization to occur, and optimum pollen germination occurs at temperatures of 27-30°C (81-86°F).

Most trees need pollinating by a cultivar with fertile pollen – not all have this.

Feeding and irrigation

Heavily cropping trees will benefit from nitrogen and potassium. In fertile soils, trees require little feeding until cropping well at 5-8 years of age. Recommended annual amounts to supply are then in the region of 30g (1oz) nitrogen + 80g (3oz) potassium per 3cm (1⅛") of trunk diameter, spread beneath the canopy area. Poor soils may require up to double these amounts, while very fertile soils may require somewhat less. Considerable amounts of nitrogen and potassium are recycled via the fallen leaves, if these are allowed to decompose beneath trees, with most nutrients released during the following growing season.

Any fertilizers should be spread early in spring. An alternative to importing nitrogen materials is to utilize nitrogen-fixing plants: interplanting chestnuts with good nitrogen fixers (e.g. alders, Elaeagnus) at the rate of 20 per cent nitrogen fixers to 80 per cent chestnuts (by canopy area) will supply all their nitrogen. Potassium accumulators like comfrey or sorrels can be used to increase potassium availability. Remember that trees with good mycorrhizal associations will require fewer nutrient additions.

Young trees are quite drought-prone and in dry summer climates irrigation is often used, at least until trees are well established. Trees under water stress vegetate and yield poorly, so drip irrigation is increasingly used in these areas.

Pruning

A little formation pruning in the first few years is usually required, to prune off low branches for access beneath trees and to keep trees pyramidal in shape. Little pruning is required later on, apart from removing dead wood. See Chapter 2, page 49 for more information on pruning sweet chestnuts.

Pests

Chestnut weevils (*Balaninus elephas*, syn. *Curculio elephas*) are serious chestnut pests in most chestnut production regions of the world. Eggs are laid in the kernels through tiny holes drilled in the shells shortly before harvest. The eggs hatch into cream-coloured grubs, which grow and tunnel through the chestnut kernels until they emerge after nut fall, overwintering in the soil at depths of 5-15cm (2-6"). In the spring some adults emerge and move into the canopies of trees, while others can remain dormant in the soil for several years. The unexpected discovery of grubs in chestnuts usually elicits a response ranging from disgust to hysteria.

The chestnut moth (*Pammene fasciana*, syn. *P. juliana*) and chestnut codling moth (*Cydia splendana*) larvae cause similar damage to weevil larvae; pheromone traps are sometimes used in the same way as for codling moth of apple. Some control of all these pests can be achieved by running poultry beneath the trees before and after nut harvesting, but bear in mind recommendations are no animals grazing for four weeks before nut harvest to reduce the risk of E. coli infections.

The oriental chestnut gall wasp (*Dryocosmus kuriphilus*) is a tiny gall-forming all-female wasp; the wasps lay their eggs in the terminal buds and the developing larvae cause shoots to become stunted. This has caused considerable damage in south-eastern USA and has spread to parts of Europe in the last decade after introduction in Italy. Some Japanese chestnut selec-

Chestnut codling moth.

Damage caused by the oriental chestnut gall moth.

tions are resistant to the wasp. There's also considerable research taking place in Asia on biological control via the use of parasitic wasps, which prey on this species.

Diseases

The two most important diseases are chestnut blight and ink disease, though there are a variety of other diseases including canker diseases.

Chestnut blight (*Cryphonectria parasitica*) is a serious fungal disease which has all but exterminated the American chestnut, and which is serious in some parts of Europe where sweet chestnut trees are affected; it is not present in Scandinavia and Great Britain. This parasitic fungus attacks the aerial parts of trees, infecting them via a natural or artificial wound (including pruning cuts, grafting, etc.) on a branch or shoot. Cankers form, eventually girdling the branch, and the upper part dies, the leaves on it drying up and reddening (appearing burned). Within a few years the main trunk is girdled, and at this stage trees often shoot from the base as if they have been coppiced. Disease spores

are normally wind-borne, but can also be carried on the feet or beaks of birds, also on insects, small mammals and slugs.

The disease is less serious on European than American chestnut because of a natural biological control due to the presence of less virulent fungal strains (called hypovirulence) and this has allowed the regrowth of many trees. The Asian chestnut species have varying degrees of resistance (Japanese chestnut, *C. crenata*, being most resistant) and these are now much used in breeding programmes with sweet chestnut to develop resistant cultivars.

There are reports that if small cankers are plastered with moist soil taken from the base of the tree, the infection doesn't spread and the cankers heal over (this has been observed with other canker diseases on other trees as well). Control is presumably due to antibiotic soil organisms. In many parts of Europe where attacks have been serious, hypovirulent strains have deliberately been introduced and allow trees to largely recover.

Ink disease (*Phytophthora cinnamomi* and *P. cambivora*) is a widely distributed fungal disease that can attack the roots of chestnut and other trees. It attacks the root bark, starting at the extremities of fine root hairs, progressing along larger and larger roots and finally attacking the trunk base. The roots cease growing and crack, releasing a flow of sap which turns black from the oxidation of tannins; the name of the disease comes from the oozing of this black liquid from the tree base in the latter stages of the attack. The attack on the root system is accompanied by the progressive death of the uppermost shoots, and little by little the whole crown. The dieback can continue progressively over several years. All chestnuts in their dormant (winter) state are susceptible to ink disease, and all are resistant when they are in active growth. In genetically resistant plants (all Japanese chestnut varieties and many hybrids like 'Marigoule' and 'Maraval') the roots recover, forming barriers of cork, and trees can withstand and recover from attacks. The main preventative measures are to plant on well-drained soils, and ensure roots are well inoculated with mycorrhizae (see Chapter 1, page 29).

Another problem exhibited by chestnuts at harvest time is a 'brown kernel' condition that appears to be some sort of internal physiological breakdown. When cut open, an affected nut shows a degree of browning and discoloration, somewhat resembling bitter pit of apples. The texture is soft and the flavour is bitter, making them inedible. Some Euro-Japanese hybrids can suffer from this, with only a proportion of nuts affected on a tree.

All species are resistant to honey fungus (*Armillaria* spp.).

Harvesting and yields

Chestnuts should be harvested promptly after

Harvested chestnuts and walnuts.

they fall (or are shaken from trees), at least every 2 days. An individual cultivar will ripen its nuts over a 7-14 day period. Hand-harvesting with Nut Wizards works well on a small scale, though there are always a few nuts that fall in-burr, so develop the 'heel-kick' method to open these. In periods of still-damp weather in autumn, it can help to knock ripe nuts down with a long pole.

Larger orchards use vacuum and sweeper harvesters and some use roll-out nets to gather the dropped nuts and burrs. Nets can also be attached to wires to keep them off the ground (leading to cleaner harvest) but this is significant work.

Sweet chestnut yields

Age of tree	Yield per tree	Yield per hectare	Yield per acre
5-8 years (*C. sativa*)	2-4kg / 4lb 6oz-9lb	140-280kg	120-250lb
5-8 years (Euro-Jap hybrids)	5-8kg / 11-18lb	345-550kg	300-480lb
8-12 years (*C. sativa*)	6-10kg / 13-22lb	415-690kg	365-600lb
8-12 years (Euro-Jap hybrids)	15-20kg / 33-44lb	1-1.4t	900-1,230lb
12-15 years plus (*C. sativa*)	15-25kg / 33-55lb	1-1.6t	900-1,400lb
12-15 years plus (Euro-Jap hybrids)	25-40kg / 55-88lb	1.6-2.8t	1,400-2,460lb

The period of harvest can occur from mid-autumn for the earliest varieties to late autumn for the latest. It is particularly important to ensure a very quick harvest of nuts after they fall, as a prolonged period on the soil in warm humid years affects the nuts (they may dry up in sun or absorb a lot of water in rain) and favours pests and pathogenic fungi.

Cropping starts after about 5 years and builds up steadily to a full crop at 15-20 years. Yields from cultivars are detailed in the table on page 235 (those for larger areas assume a tree spacing of 12 x 12m / 40 x 40') but if trees were

Sweet chestnut burrs vary in shape and length of spine between different cultivars. Clockwise from top right: burrs of 'Belle Épine', 'Marlhac', 'Marron de Goujounac', 'Marki' and 'Marigoule'.

planted at 6 x 6m / 20 x 20' and thinned out after 10 years, early yields would be near double). These are in-shell figures for fresh nuts; peeled kernel yields are about 90 per cent of these figures. Some yields from mature orchards of hybrids have reached 5t/ha (2t/acre).

Processing and storage

Fresh nuts should ideally be used or sold soon, otherwise they can start to mould. Fresh nuts stored for longer than a few weeks must be kept cool and moist.

At the time of ripening, chestnuts are about 50 per cent water. The fact that they don't quickly mould indicates that living chestnut kernel tissue has some ability to resist fungal invasion, just as living fruit and vegetables do. This 'live kernel resistance' is lost if chestnuts dry to below 35 per cent moisture, when seed viability is lost. Viability cannot be restored by rehydration. So if chestnuts are to be stored fresh, they must not be allowed to dry below 35 per cent moisture (the point at which the kernel begins to shrink away from the shell). Storage life can be enhanced by levels of high humidity, soaking, or even allowing free moisture on the nuts.

A post-harvest hot-water dip (50-68°C/122-154°F for 45-50 minutes) followed by quenching in cold water will kill any larvae inside nuts. It is difficult to identify infested chestnuts before emergence, which may take two or more weeks, though some bad nuts can be floated off (good nuts will sink). Next, carefully and slowly dry the chestnuts until the shell surfaces are free from moisture.

After water treatment, the chestnuts should be allowed to 'rest' for 3-4 days in a cool, shady building where there's enough air circulation that they do not heat but not so much that they dry. A layer of nuts up to 30cm (1') deep is acceptable. During this period, most of the free moisture will disappear from the shells and any mouldy or soft rotted chestnuts will become covered with fungal hyphae, making them easy to identify and cull (these nuts were mouldy already).

Chestnuts that are glossy and firm – i.e. fully hydrated – are relatively easy to store and transport. For long-term storage (four weeks or more), they should be kept cool (fridge temperatures) and moist with ample air circulation through the nuts. Properly stored chestnuts will begin to germinate in cold storage in midwinter.

If nuts are to be dried, then prompt harvesting followed by quick drying is another way to minimize mould losses; drying also kills any larvae in the nuts. Dry to 15 per cent moisture at 40-50°C (104-122°F): this may take 3-5 days. The kernels then become hard, grain-like and stable at room temperature. Dry to 7 per cent moisture for making flour, otherwise it will cake. (See Chapter 3, page 70 for information on drying.) Dried nuts are easy to peel (shell) and can be stored shelled or unshelled. Once shelled, mouldy nuts are easily removed. The nuts can be stored for several years.

Canning is a widely used preservation technique for chestnuts; a large variety of canned preserves, such as chestnuts in sugar syrups or chestnut purées, are available in Europe. Stabilization is reached through heating and use of ingredients that act as stabilizers: sugars and salt, which reduce water activity; or alcohol and alcoholic beverages, which inhibit microorganism growth. Sugar-rich and alcohol-steeped canned products are commonly subjected to a pasteurization heat treatment.

Frozen peeled nuts, frozen milled nuts and frozen roasted nuts, packed in aluminium or plastic bags, are now available commercially. With the prospect of small-scale peeling and milling machines becoming more available, these freezing methods will be more applicable to small growers.

Propagation

Some older cultivars are populations rather than a single genetic cultivar selection (e.g. 'Marron de Redon') and were always maintained by seed propagation. In addition, seedlings from named cultivars where the pollinator is also known to be a good cultivar will have very good potential for cropping. Seeds are not dormant and should be sown in autumn in a well-drained medium.

Most cultivars are propagated by grafting in spring or summer, and chestnut grafts require temperatures of about 21°C (70°F) to heal. European sweet chestnut (*C. sativa*) seedlings are often used as rootstocks, though some incompatibility problems can occur. Grafting on to seedlings of the cultivar to be grafted nearly always work.

Softwood cuttings can be taken from some hybrids but are difficult and require IBA rooting hormone treatment and mist. It is a difficult procedure, which works best with etiolated shoots (i.e. those grown in the dark).

Sweet chestnut seedling germinating in a seed tray 15cm (6") deep.

Cultivars

Hardiness of chestnut cultivars varies, with European sweet chestnut (*C. sativa*) selections the hardiest, followed by the hybrids, Japanese chestnut (*C. crenata*) and finally Chinese chestnut (*C. mollissima*). For challenging situations, *C. sativa* is later to leaf out, and able to be grown at higher altitudes.

The French have long divided fruit from chestnut trees into two categories: *marrons* and *châtaignes*. The categorization is dependent on whether or not the inner brown papery skin (pellicle) lies entirely or mostly on the outside of the kernel, with perhaps just a few folds into the kernel itself (*marron*); or whether it splits the kernel itself into two or more parts, called 'partitioning of the nut' (*châtaigne*). No trees produce 100 per cent of one kind or another, so a cultivar is defined as a *marron* if on average under 12 per cent of the nuts are partitioned; if, under the same conditions, a tree produces on average over 12 per cent of partitioned nuts, it is a *châtaigne*. Certain uses of chestnuts, especially commercial, demand *marrons*, which are easier to peel and process; whereas *châtaignes* are usually only for fresh eating and drying.

Hundreds of cultivars have been selected for candying, roasting, drying, flour and fresh use. Some of the best French cultivars are:

Early season: 'Marigoule', 'Vignols'
Early to mid-season: 'Marron Comballe', 'Précoce Migoule'
Mid-season: 'Bouche de Bétizac', 'Marron de Goujounac', 'Marsol'
Mid- to late season: 'Belle Épine', 'Bournette', 'Dorée de Lyon', 'Maraval', 'Marlhac'
Late season: 'Bouche Rouge', 'Maridonne'.

Pollinating sweet chestnut cultivars

Best pollinators	Fair pollinators
'Belle Épine'	'Bournette'
'Campanese'	'Maraval'
'Garinche'	'Maridonne'
'Marron de Chevanceaux'	'Marigoule'
'Marron de Goujounac'	'Marsol'
'Montagne'	'Verdale'
'Myoka'	
'Nevada'	
'Portaloune'	
'Précoce Carmeille'	
'Précoce Migoule'	
'Rousse de Nay'	
'Sauvage Marron'	
'Silver Leaf'	
'Skioka'	
'Vignols'	

Most of the good pollinating cultivars (see table left) shed pollen over a long enough period to pollinate all other varieties. These are all long-stamen types (see page 231). However, a few cultivars have very late ripening female flowers (e.g. 'Marron Comballe', 'Marron Dauphine', 'Marron d'Olargues', 'Marron du Var', 'Sardonne') and for pollination of these, stands of nearby wild trees are usually utilized.

Key to commercial sweet chestnut cultivar table

In the cultivar table on page 240, the ripening season is divided into the following six periods. Each of these covers about 8-10 days from mid- to late autumn, and 'L' and 'VL' selections can only be grown in warmer regions:

VE = very early
E = early
EM = early–mid
ML = mid–late
L = late
VL = very late

The nuts of some cultivars store very well naturally (e.g. 'Laguépie', 'Marigoule', 'Marron de Chevanceaux', 'Roussette', 'Verdale'), while others spoil very quickly after falling. A general rule is that nuts from 'ML' and 'L' ripening cultivars naturally store better than early-ripening cultivars.

Type: M = *marron*; C = *châtaigne*
Quality class: in France, Class 1 applies to the best-quality nuts; Class 2 to lesser quality
Pollinator: ✓✓ = best pollinator, ✓ = fair pollinator

Major commercial sweet chestnut cultivars: ripening season, nut type and class

Cultivar	Ripening season*						Type*	Hybrid and quality class*	Pollinator
	VE	E	EM	ML	L	VL			
'Aguyane'		✓					M	Class 2	
'Arizinca'				✓			M	Class 2	
'Bastarde'					✓		M	Class 2	
'Bastellicaciu'				✓			C	Class 2	
'Belle Épine'				✓			M	Class 1	✓✓
'Bouche de Bacon'		✓	✓				C	Class 2	
'Bouche de Bétizac'		✓					M	Hybrid, Class 1	
'Bouche de Clos'					✓	✓	M	Class 2	
'Bouche Rouge'					✓	✓	M	Class 1	
'Bournette'			✓				M	Hybrid, Class 1	✓
'Bracalla'						✓			
'Camberoune'				✓	✓		M	Class 1 / Class 2	
'Campanese'					✓		M	Class 2	✓✓
'Campbell No. 1'			✓	✓				Hybrid	
'Chalon'		✓							
'Châtaigne d'Isola'				✓	✓		C	Class 2	
'Colossal'			✓					Hybrid	
'Darlington'	✓								
'Dorée de Lyon'				✓			C	Class 2	
'Esclafarde'			✓				M	Class 2	
'Garinche'			✓				C	Class 2	✓✓
'Garrone Rosso'			✓						
'Gellatly No. 1'		✓						Hybrid	
'Gellatly No. 2'			✓					Hybrid	
'Herria'					✓		M	Class 2	
'Imperiale'				✓	✓		C	Class 2	
'Insidina'				✓			M	Class 2	
'Laguépie'				✓			C	Class 1	
'Layeroka'			✓	✓			M	Hybrid	
'Maraval'				✓			M	Hybrid, Class 2	✓✓
'Maridonne'					✓		M	Hybrid, Class 1	✓✓

Major commercial sweet chestnut cultivars: ripening season, nut type and class

Cultivar	Ripening season*						Type*	Hybrid and quality class*	Pollinator
	VE	E	EM	ML	L	VL			
'Marigoule'			✓				M	Hybrid, Class 1	✓
'Marissard'							M	Hybrid, Class 1	
'Marlhac'			✓				M	Hybrid, Class 1 / Class 2	
'Marron Comballe'					✓		M	Class 1	
'Marron Cruaud'				✓			M	Class 2	
'Marron Dauphine'				✓	✓		M	Class 1	
'Marron de Chevanceaux'				✓			M	Class 2	✓✓
'Marron de Goujounac'				✓			M	Class 1	✓✓
'Marron de Redon'				✓			C	Class 2	
'Marron d'Olargues'			✓	✓			M	Class 1	
'Marron du Var'				✓	✓		M	Class 1	
'Marrone della Val di Susa'					✓		M	Hybrid, Class 1	✓
'Marrone di Chiusa Pesio'					✓				
'Marsol'			✓				M	Hybrid, Class 1	
'Merle'			✓				M	Class 1	
'Montagne'			✓	✓			M	Class 1	✓✓
'Myoka'			✓	✓				Hybrid	✓✓
'Nevada'						✓		Hybrid	✓✓
'Nocella'				✓			M	Class 2	
'Pellegrine'				✓			M	Class 1	
'Portaloune'			✓				M	Class 2	✓✓
'Précoce Carmeille'			✓				C	Class 2	✓✓
'Précoce des Vans'		✓	✓				M	Class 2	
'Précoce Migoule'		✓					C	Hybrid, Class 1 or 2	✓✓
'Primato'	✓							Hybrid	
'Rossa'					✓		M	Class 1	
'Rousse de Nay'					✓	✓	C	Class 2	✓✓
'Roussette de Monpazier'					✓		M	Class 2	

Major commercial sweet chestnut cultivars: ripening season, nut type and class

Cultivar	Ripening season*						Type*	Hybrid and quality class*	Pollinator
	VE	E	EM	ML	L	VL			
'Sardonne'				✓			M	Class 1	
'Sauvage Marron'			✓				M	Class 2	✓✓
'Skioka'				✓				Hybrid	✓✓
'Skookum'		✓						Hybrid	
'Soulage Première'	✓						M	Class 2	
'Tichjulana'				✓	✓		M	Class 2	
'Verdale'		✓					M	Class 2	✓
'Vignols'			✓				C	Hybrid, Class 2	✓✓

* Where no details are given, this is because information is not readily available for these cultivars.

Belle Epine

Bouche de Bétizac

Bournette

Marron de Lyon

Maridonne

Marigoule

Japanese chestnut Marki

Marlhac

Marron Comballe

Marron de Goujounac

Numbo

Verdale

Vignols

Sweet chestnuts vary a lot in shape and colour.

Key to sweet chestnut cultivar table

Origin: unless otherwise described, hybrid means a European–Japanese chestnut hybrid.

Flowers:

A-stamen: without stamens, hence cannot produce any pollen. Trees sterile.

Brachy-stamen: stamen threads very small, 1-3mm (⅛") long; little pollen. Trees practically sterile.

Meso-stamen: stamen threads 3-5mm (⅛") long; little pollen. Trees practically sterile.

Long-stamen: stamen threads 5-7mm (¼") long; abundant pollen (usually fertile).

Uses: All chestnuts can be used in multiple ways, but only the main commercial uses are listed here.

Quality class: Class 1 applies to the best-quality nuts; Class 2 to those of lesser quality.

Sweet chestnut cultivars

Cultivar	Origin	Tree*	Flowers*	Nuts	Uses and quality class*
'Aguyane'	France	Erect, very vigorous, productive if pollinated well.	A-stamen	Marrons, triangular and characteristically pointed, reddish-brown but fading quickly after falling; good flavour. Ripen early over a short period. Moderate to good adaption to mechanical shelling. Moderate natural storage.	Principally for fresh consumption because of its earliness; also for canning. Class 2.
'Anderson'	USA – New Jersey	Vigorous, very productive.		Small, bright reddish-brown, downy.	
'Arizinca'	France – southern Corsica			Marrons, long-elliptical. Ripen mid-late. Very well adapted for mechanical shelling.	Very good for flour production. Class 2.
'Bartram'	USA	Vigorous, spreading, productive, has large leaves.		Small, thickly pubescent at tip, dark reddish-mahogany; good quality.	
'Bastarde'	France – south-east and Corsica			Marrons, medium-large, reddish-mahogany with black stripes, roundish-elliptical; good flavour. Ripen late over a short period. Adaptability to mechanical shelling is moderate to poor; good natural storage.	Canning. Class 2.
'Bastellicaciu'	France – southern Corsica	Very spreading.	A-stamen	Châtaignes, medium size, ripen mid-late.	Fresh consumption and flour. Class 2.

244

Sweet chestnut cultivars

Cultivar	Origin	Tree*	Flowers*	Nuts	Uses and quality class*
'Belle Épine'	France	Semi-erect, soon becoming rounded; very vigorous; very good productivity.	Long-stamen	*Marrons*, large to very large, shiny mahogany-red but fading quickly after falling, long-elliptical, thick-shelled; good flavour. Ripening mid–late within a short period, good separation from burrs. Natural storage is bad, moderate to good adaption to mechanical shelling.	Canning, confectionery, fresh. A very good, productive variety. Class 1.
'Bouche de Bacon'	France – Ardèche	Good vigour.		*Châtaignes*, medium to large, colour sometimes penetrates kernel; very good eating quality. Ripening mid-season.	Mostly used fresh. Class 2.
'Bouche de Bétizac'	Hybrid, France	Early to leaf out; very erect shape; vigorous, fruiting very rapidly, good productivity. Very adaptable. Unusual in that it retains green leaves well into late autumn.	A-stamen	*Marrons*, large, clear chestnut-red, quickly fading to dark brown, short-elliptical. Ripening early, good natural storage, very good aptitude to mechanical shelling.	Class 1.
'Bouche de Clos'	France – Ardèche			*Marrons*, medium size, ripen late to very late. Excellent natural storage.	Class 2.
'Bouche Rouge'	France – Ardèche	Very erect when young, vigorous and large-growing; regular bearing.	A-stamen and brachy-stamen	*Marrons*, medium to large, attractive shiny red, lightly ribbed, elliptical, shell medium thick; good flavour. Ripening very late over a long period, moderate falling from burrs. Slightly susceptible to chestnut codling moth and chestnut weevil. Good natural storage and adaption to mechanical shelling.	Canning, confectionery. Class 1.
'Bournette'	Hybrid, France	Semi-erect, becoming rounded; moderately vigorous; fruits rapidly (within 2-3 years); good productivity. Very adaptable.	Long-stamen	*Marrons*, medium to large, an attractive clear chestnut-brown (fades rapidly after falling), with many fine ridges and a large scar; short-elliptical, very regular, thin-shelled; good flavour. Ripening early–mid over a short period; the nuts fall well from the burrs. Very well adapted to mechanical shelling, very good natural storage.	Canning, confectionery, fresh. Class 1.
'Bourrue de Juillac'	France – Corrèze			*Châtaignes*; good eating quality. Very good natural storage.	Used fresh. Class 2.
'Bracalla'	Italy – Campania	Slow to start fruiting.		Very late ripening.	

Sweet chestnut cultivars

Cultivar	Origin	Tree*	Flowers*	Nuts	Uses and quality class*
'Camberoune'	France – Dordogne	Semi-erect, becoming rounded with age; vigorous; productive.	A-stamen	*Marrons*, small to medium, reddish (rapidly dulling after falling), triangular; good flavour. Ripen mid-late to late over a long period. Very well adapted to mechanical shelling, good natural storage.	Canning, confectionery, fresh. Class 1 or 2.
'Campanese'	France – Corsica		Long-stamen	*Marrons*, small to medium, chestnut-reddish with black stripes, elliptical with shoulders; very good flavour. Late ripening, often a short period. Badly adapted to mechanical shelling; good natural storage, but susceptible to chestnut codling moth.	Flour and canning. Class 2.
'Campbell No. 1'	Hybrid (*C. mollissima* x *C. sativa*), Canada – Ontario	Productive, reliable.		Medium size, sweet, freely falling from burrs. Mid-season ripening. Spines on burrs are finer and softer than most.	
'Canby Black'	USA – Oregon	Dwarf tree, productive.		Medium size, good flavour, shell easily.	
'Canby West'	USA – Oregon	Dwarf tree, productive.		Medium size, good flavour, shell easily.	
'Chalon'	France	Productive, precocious (early fruiting).		Small–medium, early ripening.	
'Châtaigne d'Isola'	France – south-east	Quite erect, fairly vigorous.	Brachy-stamen	*Châtaignes*, medium to large, chestnut-reddish, long-elliptical but irregular. Over 20% of nuts are partitioned. Ripen mid-late to late.	Exclusively for fresh consumption; in the past much used for flour. Class 2.
'Corrive'	France	Vigorous.		Medium size, good flavour, do not all freely fall from burrs.	
'Corson'	USA	Vigorous, spreading, very productive.		Medium size, ridged, very pubescent at tip; good quality.	
'Dager'	USA	Vigorous, spreading, productive.		Small to medium, dark brown, downy; good quality.	
'Darlington'	USA	Vigorous.		Small to medium, dark brown, striped, thickly tomentose (densely hairy) at tip, sweet; good quality. Very early ripening.	

Sweet chestnut cultivars

Cultivar	Origin	Tree*	Flowers*	Nuts	Uses and quality class*
'Dorée de Lyon' (syn. 'Marron de Lyon')	France	Leafs out late, tree is semi-erect (becoming rounded with age), moderately vigorous, moderate to good productivity. Very adaptable.	Brachy-stamen	*Châtaignes*, medium to large, an attractive shiny chestnut-red, roundish-elliptical. Over 20% of nuts are partitioned. Ripen mid–late over a short period. Good natural storage.	Exclusively for fresh eating. Productive at both low and medium altitudes. Class 2.
'Esclafarde'	France – Ardèche	Moderately vigorous, sometimes an irregular producer.		*Marrons*, large to very large, kernel flesh is sometimes yellowish. Ripen early-mid-season.	Class 2.
'Garinche'	France – Ardèche	Leafs out early, tree semi-erect, a regular producer. Very adaptable.	Long-stamen	*Châtaignes*, ripen early-mid-season. The burrs are distinctive in their small numbers of prickles.	Eaten fresh. Class 2.
'Garonne Rosso'	Italy – Campania	Good productivity, rapidly fruiting.		Ripen early-mid-season.	
'Gellatly No. 1'	Hybrid (*C. sativa* x *C. mollissima*), Canada – British Columbia	Productive.		Sweet, fall freely from burrs, which are not very prickly. Early ripening.	
'Gellatly No. 2'	Hybrid (*C. sativa* x *C. mollissima*), Canada – British Columbia	Very productive.		Good quality. Ripen early-mid season. Fall freely from burrs.	
'Grosse Noire'	France			Large, black. Good natural storage, dry well.	
'Herria'	France – Pyrénées-Atlantique		Female receptivity late, often pollinated by nearby wild trees.	*Marrons*, medium size. Late ripening.	Class 2.

Sweet chestnut cultivars

Cultivar	Origin	Tree*	Flowers*	Nuts	Uses and quality class*
'Imperiale'	France – Var	Rather erect (becoming rounded with age), moderately vigorous, hardy, regular producer.	Brachy-stamen	*Châtaignes*, large to very large, chestnut-reddish with black stripes, elliptical-roundish. Over 20% of nuts are partitioned. Ripen mid–late to late, over a long period. Good natural storage.	Fresh consumption; previously for flour. Class 2.
'Insidina'	France – Corsica		A-stamen	*Marrons*, medium size, long-elliptical, ripen mid-late. Very well adapted to mechanical shelling.	Flour and grilled *marrons*. Class 2.
'Laguépie'	France – Tarn-et-Garonne / Limousin	Semi-erect (becoming rounded with age), vigorous, good productivity, very adaptable.	Meso-stamen	*Châtaignes*, medium to large, shiny mahogany-red, elliptical-roundish, irregular; good quality. Over 25% of nuts are partitioned. Ripen mid-late, over a short period. Natural storage is moderate to good.	Exclusively for fresh consumption. Class 1.
'Layeroka'	Hybrid (*C. mollissima* x *C. sativa*), Canada – British Columbia	Vigorous, upright, pyramidal, timber-type growth. Early bearing and very productive, blight-resistant.	A-stamen	Medium size, sweet, mid-season ripening, freely falling from burrs. This variety readily propagates by layering. Sometimes overproduces, reducing nut size.	
'Lusenta'	Hybrid, Italy	Leafs out very early to early.	Mid-season.	Mediocre flavour.	
'Maraval'	Hybrid, France	Very upright, of moderate vigour, fruits rapidly, good productivity. Resistant to ink disease on its own roots. Requires very fertile soils; formation pruning is essential.	Long-stamen	*Marrons*, medium–large and very large, shiny mahogany-red with a large scar, triangular and very regular, shell medium-thick. Ripen mid–late over a short period. Natural storage is good, aptitude to mechanical shelling is mediocre.	Fresh, canning, confectionery. Propagates well by layering; prefers rich soils; doesn't like dry soils or climates. Sometimes used as a rootstock for other varieties because of its resistance to ink disease. Class 2.
'Maridonne'	Hybrid, France	Semi-erect, moderately vigorous, fruiting rapidly, good productivity.	Long-stamen	*Marrons*, large to very large, dull brown, striped, pubescent, long-elliptical. Late ripening, good natural storage, well adapted to mechanical shelling.	Class 1.

Sweet chestnut cultivars

Cultivar	Origin	Tree*	Flowers*	Nuts	Uses and quality class*
'Marigoule'	Hybrid, France	Semi-erect, becoming rounded with age; very dense and with a very straight trunk. Vigorous tree, fruiting rapidly (fourth to sixth year), very productive. Resistant to ink disease and chestnut blight.	Long-stamen	*Marrons*, medium to large and very large, shiny dark mahogany with a large scar, elliptical, thick-shelled. The flesh is very dense. Ripens early-mid-season (over a short period), with quite good release from burrs. Very well adapted to mechanical shelling, natural storage good.	Canning, confectionery. 'Marigoule' propagates well by layering and its good growth on rich soils has led it to being considered for forestry use. Sometimes used as a rootstock for other varieties because of its resistance to ink disease. Class 1.
'Marissard'	Hybrid, France	Semi-erect, of moderate vigour.	Brachy-stamen	*Marrons*, medium to large and very large, elliptical-triangular. Well adapted to mechanical shelling. Ripen early-mid-season.	Fresh and canning. Class 1.
'Marlhac'	Hybrid, France – Gironde, Dordogne, Charente	Very vigorous.	Brachy-stamen	*Marrons*, large to very large, mahogany-red, elliptical-triangular. Ripening early-mid. Natural storage is good, aptitude to mechanical shelling is good. Propagates well by layering/cuttings.	Class 1 or 2.
'Marron Comballe'	France – Ardèche, Lozère	Semi-erect, of moderate vigour, fruiting rapidly (fifth or sixth year), of good productivity.	A-stamen and brachy-stamen. Late flowering.	*Marrons*, medium to large, shiny chestnut-red with distinct black stripes, elliptical; thin-shelled, rich, sweet flavour. Ripening is late, with an irregular fall from burrs. Very well adapted to mechanical shelling, natural storage very good.	Good for fresh consumption or processing. A rugged and adaptable variety. Class 1.
'Marron Cruaud'	France – clone from the population of 'Marron de Redon' from Brittany	Semi-erect, of moderate vigour and production, fruits rapidly.	Brachy-stamen and meso-stamen	*Marrons*, medium–large, mahogany-red with ribs and distinct stripes, long-elliptical (often rounded and deformed). Very good flavour. Colour penetrates a little into the large furrows. Ripens mid-late (over a short period), with good fall from burrs.	Fresh consumption and processing. Class 2.
'Marron Dauphine'	France – Gard, Hérault, Lozère	Semi-erect (becoming rounded with age), moderately vigorous, productive if well pollinated.	A-stamen	*Marrons*, small to medium size, a clear chestnut-red, elliptical; good flavour. Ripening mid-late over a short period. Very well adapted to mechanical shelling, good natural storage but extremely susceptible to chestnut codling moth and chestnut weevil.	Mostly fresh consumption; also used for canning and the larger nuts for confectionery. Class 1.

Sweet chestnut cultivars

Cultivar	Origin	Tree*	Flowers*	Nuts	Uses and quality class*
'Marron de Chevanceaux'	France – Charente-Maritime	Semi-erect, very vigorous with good productivity.	Long-stamen	*Marrons*, medium to large, shiny chestnut-brown to mahogany-red with stripes, elliptical. Very good flavour. Ripen mid–late over a short period. Well adapted to mechanical shelling, good natural storage.	Fresh consumption and canning. Class 2.
'Marron de Goujounac'	France – western	Semi-erect (becoming rounded with age), very vigorous with very good productivity.	Long-stamen	*Marrons*, large to very large, a clear chestnut-red with black stripes, long-elliptical; good flavour. Colour penetrates well into the kernel. Ripens mid–late (over a short period). Good adaption to mechanical shelling, good natural storage.	Canning, fresh. Class 1.
'Marron de Lostange'	France – western			Medium size, shell very easily.	Class 2.
'Marron de Redon'	France – population from Brittany	Semi-erect, of moderate vigour and production, fruits rapidly.	Brachy-stamen and meso-stamen	*Châtaignes*, very large, mahogany-red with ribs and distinct stripes, long-elliptical (often rounded and deformed). Very good flavour. Nuts ripen mid–late (over a short period), with good fall from burrs.	Exclusively for fresh consumption. See also 'Marron Cruaud'. Class 2.
'Marron d'Olargues'	France – Hérault	Semi-erect, vigour moderate to vigorous, fruits rapidly (fourth or fifth year), good productivity.	Brachy-stamen	*Marrons*, medium size, chestnut-reddish with very distinct black stripes, long-elliptical. Very good flavour. The larger nuts sometimes have a high percentage of partitioning. Ripening mid- to mid-late (over a long period). Very good adaption to mechanical shelling, good natural storage.	Canning, confectionery, fresh. Prefers south-facing slopes. Class 1.
'Marron du Var'	France – south-east	Erect to semi-erect, vigorous when young but more moderate in vigour after a few years. Moderately productive.	Brachy-stamen	*Marrons*, medium to large and very large, chestnut-reddish with very distinct stripes, roundish-elliptical, thin shell. Partitioning in the large and very large nuts can be over 20% (ie putting them in the *Châtaigne* class) but is much less in medium sized nuts. Nuts fall poorly from burs. Ripening is mid-late to late, over a long period. Good adaption to mechanical shelling, moderate to good natural storage.	Canning (appearance good, flavour very good), confectionery (good), fresh (good). Renown and greatly appreciated for its good eating qualities. Class 1.
'Marrone della Val di Susa'	Hybrid, Italy – Campania			*Marrons*, medium to large and very large, chestnut-reddish with very distinct stripes, roundish-elliptical, thin shell. Very good flavour. Partitioning in large and very large nuts can be over 20% (i.e. putting them in the *châtaigne* class) but is much less in medium-size nuts. Nuts fall poorly from burrs. Ripening is mid–late to late, over a long period. Good adaption to mechanical shelling, moderate to good natural storage.	Class 1.

Sweet chestnut cultivars

Cultivar	Origin	Tree*	Flowers*	Nuts	Uses and quality class*
'Marrone di Chiusa Pesio'	Italy – Campania	Slow to start fruiting.	A-stamen type. Mid-season.	*Marrons*, small to medium, late ripening.	
'Marrone di Greve'	Italy – Campania		A-stamen		
'Marsol'	Hybrid, France	Semi-erect to erect, moderate vigour, rapidly fruiting (fourth to sixth year). Resistant to ink disease.	Long-stamen	*Marrons*, large to very large (sometimes with abnormally large burrs with over three nuts), shiny mahogany-red, very large scar, triangular to elliptical, shell medium thick. Moderate flavour. Ripen early–mid over a short period. Natural storage and adaptability to mechanical shelling are moderate to good.	Canning, possibly confectionery, fresh. 'Marsol' propagates well by layering. Sometimes used as a rootstock for other varieties because of its resistance to ink disease. Class 1.
'Merle'	France – Ardèche	Of moderate vigour.	A-stamen	*Marrons*, long-elliptical, ripening early–mid.	Mostly for fresh eating. Class 1.
'Moncur'	USA	Vigorous, spreading, very productive.		Small, light-coloured, downy.	
'Montagne'	France – Dordogne, Lot-et-Garonne	Semi-erect, vigorous, with good productivity.	Long-stamen	*Marrons*, medium to large, elliptical; good flavour. Colour penetrates into kernel. Ripens early–mid to mid–late (over a long period). Aptitude for mechanical shelling is moderate to good; natural storage is poor.	Mostly used fresh, some for canning and for whole *marron* processing. Class 1.
'Montemarano'	Italy – Campania			*Marrons*, medium to large.	*Marrons glacés*.
'Myoka'	Hybrid (*C. sativa* x *C. mollissima*), Canada – British Columbia	Upright, vigorous, timber-type growth, resistant to chestnut blight.	Long-stamen	Medium size, dark, sweet, good flavour, easy to shell. Mid-season ripening, some nuts do not fall from burrs.	
'Napoletana'	Italy – Campania	Good productivity.		Attractive, speckled.	
'Napoletanella'	Italy – Campania			Small, moderately speckled.	
'Nevada'	Hybrid (probably *C. sativa* x *C. mollissima*), USA – California	Very vigorous and upright.	Long-stamen	Large, dark, late ripening.	

Sweet chestnut cultivars

Cultivar	Origin	Tree*	Flowers*	Nuts	Uses and quality class*
'Nocella'	France – southern Corsica		A-stamen	*Marrons*, medium size, short-elliptical. Ripening mid-late, very well adapted to mechanical shelling.	Canning and flour. Class 2.
'Nouzillard'	France	Very productive.		Large, attractive.	
'Numbo'	USA	Compact, drooping, irregular cropping.		Medium to large, roundish, bright brown, striped, thinly tomentose (densely hairy). Mid-late ripening.	
'Pacora'	Italy – Campania	Very productive.		Mediocre flavour.	
'Paragon'	USA	Spreading, vigorous, very productive, narrow leaves.		Medium to large, dull brown, roundish, thickly tomentose (densely hairy), very good quality.	
'Pellegrine'	France – Ardèche, Gard, Lozère	Semi-erect to erect, very vigorous and productive.	A-stamen	*Marrons*, chestnut-reddish, elliptical-triangular. Good flavour – very aromatic. Ripen mid-late over a short period. Very well adapted to mechanical shelling, good natural storage.	Mainly for fresh consumption; also for canning. Class 1, despite its small fruits.
'Portaloune'	France – Dordogne	Erect, very vigorous, regular producer.	Long-stamen	*Marrons*, small to medium, chestnut-brown to blackish with large, dark stripes, elliptical-roundish. Moderate flavour. Ripen early-mid over a short period. Moderate to good aptitude to mechanical shelling, poor natural storage.	Fresh, canning. Class 2.
'Pourette'	France – Ardèche, Gard, Lozère	Very regular producer.		*Marrons*, small to very small.	A very old variety, known for making dried chestnuts and *crème de marrons*. Class 2.
'Précoce Carmeille'	France – Dordogne, Lot-et-Garonne	Semi-erect, moderate vigour, good productivity.	Long-stamen	*Châtaignes*, medium to large, shiny mahogany-red with distinct black stripes, elliptical. Over 20% of nuts are partitioned. Ripens early-mid over a short period.	Exclusively for fresh consumption. Class 2.
'Précoce des Vans'	France – Ardèche, Gard, Lozère	Of moderate vigour.	Brachy-stamen	*Marrons*, small to medium size, chestnut-reddish, elliptical. Colour penetrates into the kernel. Ripens very early to early, over a short period.	Exclusively for fresh eating because of its earliness. Class 2.

Sweet chestnut cultivars

Cultivar	Origin	Tree*	Flowers*	Nuts	Uses and quality class*
'Précoce Migoule'	Hybrid, France	Semi-erect and sparsely branched, moderately vigorous, good productivity, fruits very rapidly (within 2-3 years).	Long-stamen	*Châtaignes*, medium (to large), an attractive clear mahogany, elliptical-triangular with a large scar and medium-thick shell. Over 20% of nuts are partitioned. Quite good release from burrs. Ripening is early (over a short period). Natural storage is moderately good, good aptitude to mechanical shelling.	Used mainly for fresh consumption on account of its early ripening and good-sized fruits. Propagates well by layering. Class 1 or 2.
'Primato'	Hybrid, Italy – Campania	Starts fruiting very early, very productive.	Very early.	Ripen extremely early.	
'Prolific'	USA	Vigorous, spreading, very productive, reliable, leaves small and narrow.		Medium size, dark brown; good quality. Early ripening.	
'Quercy'	France	Productive and early bearing.		Medium-sized, good quality.	
'Radulacciu'	France – southern Corsica			*Châtaignes*.	Fresh consumption and flour. Class 2.
'Ridgely'	USA	Vigorous, spreading, very productive.		Medium size, dark, moderately downy; very good quality and flavour.	
'Rossa'	France – Corsica	Semi-erect, good vigour and productivity.	A-stamen	*Marrons*, medium to large (very regular), attractive clear shiny mahogany with black ridges, shape irregular (round to elliptical-triangular). Very good quality. Colour penetrates well into the kernel. Late ripening (over a short period). Good aptitude for mechanical shelling, natural storage good.	Fresh consumption or processing; extensively used in Corsica. A clone of this population ('Zalana') is noted for the non-penetration of colour into the kernel, and for its ease of mechanical shelling. Class 1.
'Rougières'	France – Dordogne			*Châtaignes*, medium size, elliptical-roundish. Good quality.	Fresh eating. Class 2.

Sweet chestnut cultivars

Cultivar	Origin	Tree*	Flowers*	Nuts	Uses and quality class*
'Rousse de Nay'	France – Pyrenees	Relatively erect, moderately vigorous, good productivity.	Long-stamen	*Châtaignes*, medium (to large), an attractive shiny mahogany-red with distinct black stripes, elliptical and irregular. Over 20% of nuts are partitioned. Ripen late to very late, over a long period. Good natural storage.	Exclusively for fresh consumption. Class 2.
'Roussette de Monpazier'	France – Lot, Lot-et-Garonne, Dordogne			*Marrons*, elliptical-triangular, late ripening.	Excellent for fresh eating because of very easily removed shell. Class 2.
'Sardonne'	Italy	Quite erect, of moderate vigour, with good productivity if well pollinated.	A-stamen	*Marrons*, medium to large, reddish with black stripes, elliptical, thin-shelled. Good flavour. Ripen mid–late over a short period. Good aptitude to mechanical shelling, natural storage very good.	Canning. Class 1.
'Sauvage Marron'	France – Lot, Corrèze		Long-stamen	*Marrons*, ripen early–mid.	Class 2.
'Scott'	USA	Open, spreading, very productive.		Small to medium, slightly pointed, glossy dark brown. Relatively free from weevil attack.	
'Settlemeier'	Hybrid (*C. mollissima* x *C. sativa*), USA	Vigorous, spreading, rounded.		Medium to large.	
'Silver Leaf'	Hybrid, USA	Productive, reliable. Undersides of leaves turn silvery-grey as nuts ripen.	Long-stamen	Medium to large, sweet, easily shelled.	
'Simpson'	Hybrid (*C. mollissima* x *C. sativa*), USA	Very productive.		Large.	
'Skioka'	Hybrid (*C. sativa* x *C. mollissima*), Canada – British Columbia	Vigorous, upright, timber-type growth.	Long-stamen	Medium size, dark, sweet, mid–late ripening. Not all freely falling from burrs.	

Sweet chestnut cultivars

Cultivar	Origin	Tree*	Flowers*	Nuts	Uses and quality class*
'Skookum'	Hybrid (*C. sativa* x *C. mollissima*), USA	Vigorous, upright, very productive and reliable, moderately early bearing, leaves drop quite early.		Medium size, shiny, attractive, sweet, fall freely from burrs. Ripen early.	
'Soulage Première'	France – Gard, Hérault	Erect, very vigorous, productive.	Brachy-stamen	*Marrons*, medium to large, chestnut-reddish with distinct black stripes, elliptical. Ripen very early over a short period.	Usually for fresh consumption because of its earliness; also for canning. Class 2.
'Styer'	USA	Very vigorous, upright, large leaves.		Small, dark brown, striped, pointed, tomentose (densely hairy) at the tip.	
'Tichjulana'	France – Corsica			*Marrons*, small to medium, reddish with black stripes, round; moderate flavour. Ripen mid-late to late over a long period. Adaption to mechanical shelling moderate; natural storage good.	Fresh, flour, canning. Class 2.
'Tricciuda'	France – Corsica		A-stamen	*Marrons*, medium to large, short-elliptical. Very well adapted to mechanical shelling.	Class 2.
'Verdale'	France – Cantal, Dordogne, Lot; Italy – 'Verdole' and Switzerland – 'Verdesa'	Semi-erect, vigorous, with good productivity.	Long-stamen	*Marrons*, small to medium, dark chestnut-brown with blackish stripes, elliptical. Early ripening. Natural storage very good.	Class 2.
'Vignols'	Hybrid, France	Rather erect but soon becomes rounded with age. Very vigorous and moderately productive.	Long-stamen	*Châtaignes*, large to very large, dark reddish, elliptical-triangular, thick-shelled. Colour penetrates somewhat into the kernel. Over 20% of nuts are partitioned. Ripen early-mid over a short period. Moderate natural storage.	Fresh consumption. Class 2.

* Where no details are given, this is because information is not readily available for these cultivars.

Cultivars of American /Chinese/Japanese chestnut origin

Cultivars of the American chestnut (*C. dentata*) are still sometimes grown in North America outside the natural range of the species (e.g. western USA), where they are often safe from chestnut blight; but few, if any, are commercially available now. American chestnuts bore smaller nuts than other species, but the nuts were very sweet and rich eaten raw. More commonly planted are various hybrids of American chestnut and the Chinese and Japanese chestnuts, several of which are blight-resistant and are grown commercially.

Chinese chestnut (*C. mollissima*) cultivars are often grown in North America. The Chinese chestnut is resistant, but not immune, to chestnut blight (cankers usually heal) and is highly resistant to ink disease. However, it is susceptible to honey fungus (*Armillaria* spp.), while other chestnuts are not. It requires slightly warmer conditions than sweet chestnut (*C. sativa*). The nuts are sweeter and richer than most European chestnuts.

Japanese chestnut (*C. crenata*) varieties are commonly grown in North America, southern Europe and other parts of the world. The Japanese chestnut is also highly resistant to blight (though slightly less so than the Chinese chestnut) and is highly resistant to ink disease. It requires slightly warmer conditions than sweet chestnut (*C. sativa*). The nuts are coarser and starchier than those of other chestnuts.

Recommended American, Chinese and Japanese chestnut cultivars

Cultivar	Origin	Tree*	Flowers*	Nuts
'Abundance'	*C. mollissima*, USA			Rich brown, good flavour. Easily shelled.
'Alachua'	Hybrid (*C. dentata* x *C. mollissima*), USA	Upright, vigorous, productive. Immune to chestnut blight.		Medium size, sweet, easily shelled.
'Alpha'	C. crenata, USA	Upright, very vigorous, productive.		Small to medium, dark, fair quality. Ripens very early.
'Appalachia'	Hybrid (*C. dentata* x *C. mollissima*), USA	Upright, large, vigorous early bearing (2-4 years). Immune to blight.		Medium size, sweet, easily shelled.
'Armstrong'	Hybrid (*C. mollissima* x *C. dentata*), USA	Very upright.		Medium size, very sweet.
'AU-Cropper'	*C. mollissima*, USA – Alabama	Very productive.		Small to medium, dark chocolate-brown, glossy, attractive. Ripen mid-season. Fall freely from burrs.
'AU-Homestead'	*C. mollissima*, USA – Alabama	Very productive.		Small to medium, very dark chocolate-brown. Late ripening over long period. Fall moderately well from burrs.

Recommended American, Chinese and Japanese chestnut cultivars

Cultivar	Origin	Tree*	Flowers*	Nuts
'AU-Leader'	*C. mollissima*, USA – Alabama	Very productive.		Medium size, dark chocolate-brown, glossy and attractive, Mid-season ripening. Fall freely from burrs.
'Beta'	*C. crenata*, USA			Small, light brown, good quality. Ripen early.
'Biddle'	*C. crenata*, USA	Rounded, vigorous, reliable.		Medium size, bright brown, broad, downy, fair quality. Ripen mid-season.
'Bill's Earliest'	Hybrid (*C. mollissima* – other species unknown), Canada – British Columbia			Very early ripening. Fall freely from burrs.
'Black'	*C. crenata*, USA	Rounded, dense, vigorous, productive.		Small to medium, irregular, dark brown, good quality. Early ripening.
'Black Beauty'	*C. mollissima*, USA	Moderately productive.		Small to medium. Early ripening. Fall freely from burrs.
'Boone'	Hybrid (*C. dentata* x *C. crenata*), USA	Vigorous, productive, precocious.		Medium to large, light brown, downy, good quality. Early ripening.
'Carolina'	Hybrid (*C. dentata* x *C. mollissima*), USA	Vigorous, spreading, productive, early bearing (2-4 years). Immune to chestnut blight.		Large to very large, shiny chocolate-brown, very sweet.
'Carpentar'	Hybrid (*C. dentata* x *C. mollissima*), USA	Vigorous, productive. Immune to chestnut blight.	Long-stamen	Medium size, reddish-brown, sweet.
'Carr'	*C. mollissima*, USA			Very small, sweet, very downy.
'Coe'	*C. crenata*, USA – California	Upright, somewhat spreading with age.		
'Colossal'	Hybrid (*C. crenata*, *C. mollissima*, *C. sativa*), USA – California	Vigorous, productive.		Medium size, very sweet.
'Crane'	*C. mollissima*, USA	Upright, very early to bear (2-3 years), resistant to chestnut blight.	Long-stamen	Medium to large, sweet, easy to shell, medium quality. Ripen mid-season. Fall freely from burrs. Dry and store well.
'Douglass'	Hybrid (*C. mollissima* x *C. dentata*), USA – New York	Upright.		Medium to large, dark cherry-red, good flavour. The burrs are exceptionally large. Natural storage is very good.
'Douglass #1'	Hybrid (*C. mollissima* x *C. dentata*), USA – New York	Productive, blight-resistant.		Small, sweet flavour.
'Douglass #1A'	Hybrid (*C. mollissima* x *C. dentata*), USA – New York	Productive, blight-resistant.		Small, good flavour.

Recommended American, Chinese and Japanese chestnut cultivars

Cultivar	Origin	Tree*	Flowers*	Nuts
'Douglass #2'	Hybrid (*C. mollissima* x *C. dentata*), USA – New York			Medium size, very good sweet flavour. Early ripening.
'Douglass Manchurian'	Hybrid (*C. mollissima* x *C. dentata*), USA – New York	Very blight-resistant.		Medium size, good quality.
'Dulaney'	*C. dentata*, USA			Medium size, good quality.
'Dunstan'	Hybrid (*C. dentata* x *C. mollissima*), USA	Vigorous, precocious (bears in 3-4 years), highly blight-resistant.		Easy to shell, very good flavour.
'Eaton'	Hybrid (*C. mollissima*, *C. crenata*, *C. dentata*), USA	Like *C. mollissima*, early bearing.		Medium size, sweet, good flavour. Early ripening.
'Ederra'	*C. crenata*, France	Fruits quickly.		*Marrons*, large, very sweet. Ripen early–mid-season.
'Etter'	Hybrid (*C. dentata* x *C. mollissima*), USA	Upright, vigorous, blight-resistant.		
'Felton'	*C. crenata*, USA	Rounded, moderate productivity.		Small, dark brown, good quality. Early ripening.
'Ford's Sweet'	Hybrid (*C. dentata*, *C. crenata*, *C. mollissima*), USA	Upright, vigorous, timber-type growth; heavy bearing, early to start bearing (3-4 years).		Small, sweet.
'Griffin'	*C. dentata*, USA			Medium size, very downy, good quality.
'Grimo 142Q'	*C. mollissima* or hybrid, Canada – Ontario	Very productive.		Medium size.
'Grimo 150Y'	*C. mollissima* or hybrid, Canada – Ontario	Vigorous, moderately productive, , good resistance to chestnut blight.		Large, sweet, easily peeled.
'Hale'	*C. crenata*, USA – California	Very early bearing.		Medium to large, dark brown, good quality.
'Hathaway'	*C. dentata*, USA – Michigan	Productive, reliable.		Medium size, light-coloured, sweet.
'Henry VIII'	*C. mollissima*, USA	Vigorous at first, slowing once bearing begins.		Medium size, shiny mahogany, with a very good flavour. Kernel is yellowish.
'Heritage'	Hybrid (*C. dentata* x *C. mollissima*), USA	Very vigorous and erect, precocious (bears in 2-4 years), straight-trunked with a good timber form. Low productivity, blight-resistant.		
'Hobson'	*C. mollissima*, USA – Georgia	Productive, reliable, early bearing.		

Recommended American, Chinese and Japanese chestnut cultivars

Cultivar	Origin	Tree*	Flowers*	Nuts
'Honan'	*C. mollissima*, USA - Oregon			Small, elongated, chocolate-brown, very sweet. Ripen late, over a short period. Nuts fall well from burrs.
'Ipharra'	*C. crenata*, France	Quickly fruiting, very productive.		Small, very sweet.
'Jersey Gem'	*C. mollissima*, USA	Rounded, moderately vigorous, reliable, very productive.		Small, fair quality.
'Kent'	*C. crenata*, USA - New Jersey	Rounded, productive, early bearing.		Small to medium, dark, good quality. Ripen very early.
'Kerr'	*C. crenata*, USA - New Jersey	Vigorous, rounded, very productive.		Small to medium, dark brown, broad. Early ripening.
'Ketcham'	*C. dentata*, USA – New York	Vigorous, productive.		Small, oblong, downy, sweet.
'Killen'	*C. crenata*, USA – New Jersey	Upright, open, moderately vigorous and productive.		Large, light brown, slightly ridged, good quality. Ripen mid-season.
'Kuling'	*C. mollissima*, USA – Georgia	Quite upright, vigorous,		Small to medium. Nuts freely fall from burrs. Good natural storage.
'Kungki'	*C. mollissima*, USA	Vigorous, rapidly fruiting (3 years).		Medium size, very attractive.
'Linden'	Hybrid, USA		Long-stamen	Medium size.
'Lockwood'	Hybrid (*C. dentata* x *C. mollissima*), USA	Vigorous, productive, reasonable blight resistance.		Very large, sweet, easily peeled.
'Manoka'	*C. mollissima*, Canada – British Columbia	Upright, timber-type growth, productive.		Medium size, dark brown, easily shelled, good flavour. Yellow kernels.
'Marki'	*C. crenata*, France	Semi-upright tree, quickly fruits.		*Marrons*, large to very large, long and narrow, very sweet. Ripen early–mid-season.
'Martin'	*C. crenata*, USA – New Jersey	Vigorous, open, productive.		Medium to large, bright reddish-brown, broad. Ripen mid-season.
'McFarland'	*C. crenata*, USA - California	Spreading, very productive.		Medium size, good quality. Early ripening.
'Meiling'	*C. mollissima* or hybrid, USA	Quite upright, heavy and early bearing.		Small to medium size, good flavour, early ripening. Good storage qualities.
'Mossbarger'	Hybrid involving *C. mollissima*, USA	Very productive.		Medium size, sweet, good natural storage.
'Murrell'	*C. dentata*, USA			Medium to large, good flavour.
'Nanking'	*C. mollissima*, USA	Spreading, vigorous, very early to bear (2-3 years), reliable and productive.		Medium size, dark tan, some split on falling. Ripen mid-late season.

Recommended American, Chinese and Japanese chestnut cultivars

Cultivar	Origin	Tree*	Flowers*	Nuts
'Nevada'	Hybrid, USA	Selected as a pollinator for 'Colossal'.	Long-stamen	Small to medium, dark brown, very sweet.
'Orrin'	C. mollissima, USA	Erect, low vigour, early bearing.		Medium to large, dark mahogany with a light scar, good flavour, good natural storage. Easy to shell.
'Otto'	C. dentata, USA			Medium size, oblong, very downy at tip, rich and very sweet.
'Parry'	C. crenata, USA	Open, spreading, moderate vigour, large leaves.		Large, dark brown, ridged, fair quality.
'Penoka'	C. mollissima, Canada – British Columbia	Upright, timber-type growth, productive, reliable.		Medium to large, good flavour, easily shelled.
'Qing'	C. mollissima, USA – Ohio	Blight-resistant.		Large, sweet, easily peeled.
'Reliance'	C. crenata, USA – New Jersey	Dwarfish, spreading, drooping, very productive and early bearing; may overbear.		Small to medium, light brown, long, ridged, fair quality. Ripen mid-season.
'Revival'	Hybrid (C. dentata x C. mollissima), USA	Vigorous, upright, spreading in upper canopy. Precocious (bears in 2-4 years), heavy and reliable annual cropper. Blight-resistant.		Large, dark reddish-brown, sweet, easily shelled. Ripen over a short period.
'Rochester'	C. dentata, USA	Very vigorous and productive.		Small to medium, dull brown, rounded, downy at the tip, excellent quality. Ripen late.
'Sleeping Giant'	Hybrid (C. mollissima, C. crenata, C. dentata), USA	Vigorous, somewhat spreading, a heavy and reliable bearer, with large leaves. Blight-resistant.		Small to medium size, attractive, good quality, easily shelled.
'Success'	C. crenata, USA – New Jersey	Upright, productive.		Medium to large.
'Superb'	C. crenata, USA – New Jersey	Vigorous, very productive.		Medium size, broad, fair quality. Early ripening.
'Wards'	Hybrid (C. dentata x C. mollissima) USA	Very vigorous.		Small, sweet, yellow kernels. Ripen early–mid-season.
'Watson'	C. dentata, USA			Small to medium, slightly downy, good quality.
'Willamette'	Hybrid (C. dentata x C. mollissima), USA	Of moderate vigour, semi-erect, very productive, precocious (bears in 2-4 years). Blight-resistant.		Very large, reddish-brown, sweet, easily shelled.

* Where no details are given, this is because information is not readily available for these cultivars.

TRAZELS

(*Corylus* spp.)

ZONE 4-5, H7

Trazels are hybrids of European hazel (*Corylus avellana*) and other hazel species with tree form, which were developed by the famous nut breeder James U. Gellatly of British Columbia, Canada, from the 1930s to the 1960s. He also coined the term 'trazel'. The objectives were to achieve selections that grew as non-suckering, single-stemmed trees but that also had good-quality large nuts.

Turkish hazel (*Corylus colurna*) is one of the parents of trazels

Another amateur nut breeder, C. Farris of Michigan, USA, continued Gellatly's work, concentrating partly on resistance to eastern filbert blight.

Several species were used in the hybridizing, including most commonly Turkish hazel (*C. colurna*), but also Chinese hazel (*C. chinensis*) and Indian tree hazel (*C. jacquemontii*). These species themselves grow as medium or large single-stemmed trees, but bear only small nuts.

Trazel trees resemble European hazel in leaf form. They grow into larger, more upright trees than hazel, the leafy bracts around the nuts are usually frilly, and the nuts are intermediate in character, depending on the parent. Their lifespan is similar to hazel – around 80 years.

Uses

The nutritional composition of trazel nuts is akin to hazelnuts (see Appendix 1) and they can be used in a similar way (see page 149). Most trazel selections have very good kernel quality – light colour, no shrinkage on drying, and a fine crisp texture.

Developing nuts on the trazels (clockwise from top left) 'Chinoka', 'Faroka', 'Morrisoka' and 'Grand Traverse'.

Cultivation

Trazels generally prefer a slightly warmer climate and sunnier position than hazels, but like hazels they crop better in a sheltered location because they flower early. Trazels often grow larger than hazels and are destined to become medium-size trees in time.

Rootstocks and soils

Soil preferences and planting arrangements are as for hazel (see page 150). If grafted on to Turkish hazel, trazel trees will be rather more drought-tolerant than hazels and may grow slightly larger, so in this case increase planting distances a little.

Pollination

Trazels will need cross-pollinating by other trazels or by hazels.

Feeding and irrigation

As for hazel (see page 151).

Pruning

Trazels are grown as single-stemmed trees, so low branches are gradually pruned off as trees grow. Little other pruning is required unless grafted on to hazel rootstock: when rootstock suckers will need to be cut out each year.

Pests

Squirrels and other pests are as for hazel (see page 152). Turkish hazel is very resistant to bud mites and this may have been inherited in some trazels.

Diseases

As for hazel (see page 153). Turkish hazel is very resistant to eastern filbert blight and this may have been inherited in some trazels.

Harvesting and yields

Harvest from the ground when the nuts fall, as for hazels (see page 153). Yields may be more than those for hazels, once trees become larger.

Processing and storage

As for hazel (see page 154).

Propagation

Named cultivars are grafted on to European hazel (*C. avellana*) or Turkish hazel (*C. colurna*) rootstocks, the latter likely leading to larger trees. Good warmth is needed at the graft union for grafts to heal, so a hot graft pipe or similar technique is useful.

Cultivars

Only a few trazel cultivars are available commercially, so you may be limited to what you can find (see table). North American growers may be especially interested in recent native selections that are resistant to eastern filbert blight.

Recommended trazel cultivars

Cultivar	Comments
'Chinoka'	Hybrid of *C. avellana* and *C. colurna*. Nuts medium size, conical.
'Eastoka'	Hybrid of *C. avellana* and *C. colurna*. An annual heavy cropper of fair-sized nuts. It is also a heavy producer of catkins annually, making it a good pollinator. The nuts do not fall from the husk – bunches of 5-6 nuts fall as a cluster and mostly remain in the husk.
'Faroka' (syn. 'Freeoka')	Hybrid of *C. avellana* and *C. colurna*. Nuts medium size, very fine quality. Tree a good producer.
'Grand Traverse'	'Faroka' x *C. avellana*, resistant to eastern filbert blight. Nuts nearly round, brown; 51% kernel by weight; ripens with hazelnut cultivar 'Ennis'; kernel plump, clean, excellent flavour. Tree vigorous, productive, not precocious (early fruiting), winter-hardy. Resistant to big bud mites.
'Grimo 186M'	More recent hybrid of 'Faroka'. Nuts medium size, oval. Upright tree, moderate bearer, resistant to eastern filbert blight and bud mite.
'Grimo 208D'	More recent hybrid of 'Faroka'. Nuts medium size, oval. Upright tree, moderate bearer, resistant to eastern filbert blight and bud mite.
'Indoka'	Hybrid of *C. avellana* and *C. jacquemontii*.
'Karloka'	Hybrid of *C. avellana* and *C. colurna*.
'Laroka'	Hybrid of *C. avellana* and *C. colurna*. Vigorous tree with very large dark green leaves and good nut size. It sets a fair crop in clusters up to 5-6 nuts, which fall free of the husk when mature.
'Lisa'	'Grand Traverse' x *C. avellana*. Resistant to eastern filbert blight.
'Morrisoka'	Hybrid of *C. avellana* and *C. colurna*. Nuts are large, high quality, growing in clusters of 4-6 nuts that are free-falling when mature. Tree with large dark green leaves that drop late.

WALNUT

(Juglans regia)

ZONE 4-5, H7

For most people, walnut needs little introduction, having been used for food in Europe, the Caucasus and the Middle East for thousands of years. Before cultivation in Ancient Greece, walnut was cultivated extensively for several thousand years in Persia (Iran). The natural range is from the Carpathian Mountains in Poland to the Middle East and the Himalayas.

Walnuts tree in a commercial orchard.

Walnut is a fast-growing, broad-headed deciduous tree reaching up to 18-30m (60-100') tall and 12-18m (40-60') wide with irregular branches. The bark of upper branches is smooth and ash-coloured; older bark is grey and becomes deeply fissured in an irregular pattern. Trees can be long-lived – 250 years or more.

Leaves are light green, compound, 20-30cm (8-12") long on vigorous young shoots, usually with 5-7 leaflets. The leaves are strongly scented, especially when crushed. They are high in nitrogen (2.5-3 per cent), phosphorus (0.19-0.25 per cent) and potash (1-1.3 per cent), and decompose quickly after leaf fall, speeding nutrient cycling. In the autumn they turn yellowish-brown before falling. Bud break occurs in mid-spring, depending on the cultivar, and leaf fall occurs in late autumn.

Female flowers in spring.

The flowers open before or around the same time as the leaves (so late-leafing cultivars are generally also late flowering and more likely to escape late frost damage). Flower initiation is probably dependent on suitable conditions the previous summer. Male flowers are very numerous, borne on slender pendulous catkins 5-10cm (2-4") long, which grow from lateral buds on wood of the previous season's growth. Female flowers are tiny, in small, short spikes, and are mostly borne terminally on the current season's growth; a few precocious (early-flowering) cultivars will also bear female flowers on lateral growth.

Walnut produces both male and female flowers but it is common for male and female flowers to mature at different times, so most trees are not very self-fertile.

The fruit, developing from the female flowers, consist of a single nut surrounded by a thin, fleshy husk 37-62mm (1½-2½") across, and mostly produced in clusters of one to three. The husks split when ripe in autumn and the nuts drop to the ground.

There are several naturally occurring forms (from which a number of cultivars have been selected) and two of these are of more interest to nut growers:

* f. *macrocarpa*, known as 'ban-nut', 'claw-nut' or *noyer à bijoux*, has nuts about twice the ordinary size, and kernels 30-60 per cent larger than normal. Several cultivars have been selected; unfortunately, most seem quite disease-susceptible.

* f. *rubra* has blood-red-coloured pellicles (skins around the kernels), the dried kernels appearing pinky-red. Several cultivars have been derived.

Uses

The nuts are highly nutritious (see Appendix 1) and have numerous health benefits. They can be eaten on their own (raw, toasted or pickled) or cooked in many savoury and sweet dishes, and in preserves.

Oil can be pressed from the ripe nuts (around 50 per cent by weight of kernels). The oil can be used raw in salads, for cooking or as a butter substitute. The oil should be kept in a cool, dark place to prolong storage.

The finely ground shells are used in the stuffing of agnolotti pasta.

Secondary uses of walnut

The sap of the walnut tree is edible, tapped in the same way as that of sugar maple. You wouldn't normally want to tap your good nut-producing trees, as tapping reduces the vigorous growth and yields from a tree. But if, for example, you were going to thin out trees, then you could tap the ones to be thinned for a few years before removing them.

The timber is very stable, scarcely warps or checks at all, and after proper seasoning swells very little. The wood is uniform and straight-grained, fairly durable, slightly coarse (silky) in texture so easily held, strong, of medium density and can withstand considerable shock. It is easy to work and holds metal parts with little wear or risk of splitting. The heartwood is mottled with brown, chocolate, black and pale purple colours intermingled. Some of the most attractive wood comes from the root crown area, from which fine burr walnut veneers can be obtained. The timber is mostly used for veneers; also for rifle butts, high-class joinery, plywood and wooden bowls. It makes excellent firewood. In former times it was used for the wheels and bodies of coaches.

Several parts of the tree have traditional medicinal uses, mostly due to the juglone content, including:

* The leaves have many properties, for example astringent, detergent, laxative, and are used for the treatment of skin fungal diseases. Leaves should be picked in midsummer in fine weather, and dried quickly in a shady, warm, well-ventilated place.
* Male catkins are made into a broth and used to treat of coughs and vertigo.
* The nuts are antilithic (help prevent and remove kidney stones), diuretic and stimulant.
* The cotyledons (embryonic leaves) have been used in the treatment of cancer. Some extracts from the plant have shown anti-cancer activity.
* The juice of the green husks, boiled with honey, is a good gargle for sore throats.
* The nut oil expels parasitc worms and can be used for colic and skin diseases.
* The root bark is astringent. The husks and shells are sudorific (induce sweating), especially when green.
* The plant is used in Bach Flower Remedies.

The husks (and leaves to an extent) are well known as dye materials – something you will know if you have ever hand-harvested walnuts! The green husks can be boiled to produce a dark yellow dye, while the leaves and mature husks yield a brown dye used on wool (with alum or no mordant) and to stain skin for a week or so. The green husks are also a good source of tannin, and have been used as a wood stain and preservative. The leaves and husks can be harvested and dried for later use. A golden-brown dye is obtained from the catkins in early summer. It doesn't require a mordant.

Walnut oil has been used for making varnishes, polishing wood, in soaps and as a lamp oil.

Juglone in the leaves is also insecticidal, and they act as an insect repellent; in former times, horses were rested beneath walnuts to relieve them of insect irritation. Leaves can be crushed for greater effect. They also show some herbicidal effects.

The walnuts grow inside green husks. The shape of the husk reflects that of the nut inside: (clockwise from top left) 'Buccaneer', 'Chandler', 'Meylanaise', 'Ronde de Montignac'.

Cultivation

The best sites are sheltered, sunny sites, mid-slope on slight south- or south-west facing slopes, free from unseasonable frosts – because young growth and flowers are damaged by even short spells below -2°C (28°F). Full overhead light and ample side light is required for trees to grow with large rounded crowns, enabling maximum nut production. The southern limit for walnut cultivation is delineated by the chilling requirements (period of cold winter temperatures) of 500-1,500 hours, and the northern limit by length of growing season and spring frost risk.

A pollinating tree should be no further than eight trees away, i.e. around 100m (330'), though in humid climates this minimum should be reduced to around 50m (160').

Aim to space mature trees at 9-15m (30-50'), with most cultivars preferring the larger end of this range. As with other (ultimately) large nut trees, the generous spacing means there's an opportunity to intercrop with more walnut or other fruit or nut trees, or lower crops (see Chapter 1, page 34). Intercrop trees will need removing after 10-15 years to allow the walnuts to become full sized. Dwarf cultivars can be planted at 4m (13') spacing.

Some orchards in warm regions are planted in a hedgerow system, for example with rows 7m (23') apart and trees 5m (16') in the row. The fruiting faces are machine-trimmed by hedging machines every 3-4 years to allow fruiting spurs to regenerate. It is unknown if this system is appropriate to cooler regions.

Plant trees when dormant in winter. Walnuts have large taproots and bare-rooted trees will often have larger roots than other fruit trees and will need a larger planting hole. Staking isn't usually required, unless trees are pot-grown. Protect against rabbits and other grazing animals. In heavy soils, planting on a slight mound can be beneficial in increasing drainage around the crown.

Mulch trees well to a diameter of at least 1m (3'), as young walnuts are very susceptible to grass competition. Aerial growth in the first few years is slow, as trees concentrate on growing roots. After about 5 years, trees should grow at 20-40cm (8-16") per year, and 10 years after planting they should be 3-4m (10-13') high and wide.

Rootstocks and soils

In Europe, cultivars are usually grafted on to seedling walnut (*J. regia*) rootstocks. This produces good trees with an excellent graft union, which are very resistant to crown rot; but the roots are not resistant to honey fungus (*Armillaria* spp.), and moderately resistant to root knot nematodes (*Meloidogyne* spp.).

In parts of North America, especially California, the following are often used as rootstocks:

* Black walnut (*J. nigra*): resistant to crown gall.
* Northern California walnut (*J. hindsii*): resistant to honey fungus and root knot nematode.
* The 'Paradox' hybrid *J. hindsii* x *regia*: used in heavy or poor soils and tolerant of wet conditions; resistant to crown rot (*Phythophthora* spp.) and root lesion nematodes.

In recent years, three clonal Paradox rootstocks have become available; Vlach, RX1 and VX211. Each clonal Paradox rootstock has specific beneficial characteristics. For example, Vlach is vigorous and may be moderately tolerant of crown gall compared with Paradox seedlings. VX211 is very vigorous and shows some tolerance to nematodes. RX1 is a less vigorous rootstock but has shown moderate resistance to *Phytophthora citricola* and high resistance to *P. cinnimomi*. The risk of using any rootstock other than *Juglans regia* is blackline disorder (see 'Diseases', page 274).

The soil should be moist, well drained and fertile, ideally with a pH of between 6 and 7 (though pH 4.5 to 8.3 is tolerated). A deep medium loam is preferred, without a compacted soil layer or high water table. Walnut does not like light, sandy soils or very heavy soils.

Pollination

Walnut flowers in mid- to late spring, and is wind-pollinated.

As with many trees, male and female flowers are borne on the same tree, but there is often insufficient overlap of pollen shedding and female flower receptivity (female flower readiness), meaning few walnuts are very self-fertile. Cultivars are protandrous if pollen is shed before female flowers are receptive, and protogynous if pollen is shed after. All commercial walnut

Female and male walnut flowers.

orchards use more than one cultivar and ensure a mix of protandrous and protogynous types to get reliable pollination of all trees. Smaller plantings should do the same, unless there is only room for a single tree, when one of the few highly self-fertile cultivars should be chosen.

Flowers are severely damaged even by light spring frosts, hence planting late-flowering selections may be prudent in many areas.

Feeding and irrigation

The nutrient requirements for walnuts that are cropping well are similar to those for sweet chestnuts, i.e. nitrogen and potassium. Walnuts respond well to nitrogen, but excess nitrogen makes them much more susceptible to walnut blight, so do not overfeed. Normal recommendations are to supply 50g (2oz) nitrogen + 50-80g (2-3oz) potassium per 3cm (1⅛") of trunk diameter per year, spread beneath the canopy area. Considerable amounts of nitrogen and potassium are recycled via the fallen leaves, if these are allowed to decompose beneath trees.

Any fertilizers should be spread early in spring. The nitrogen component shouldn't be long-lasting – the object is to have a high nitrogen level in the spring to speed growth, but a low nitrogen level in late summer and autumn to slow growth and aid the hardening-off of wood. Hence, no fertilizer should be applied after midsummer.

Interplanting nitrogen fixers and potassium accumulators can supply many or all the nutrients needed (see 'Sweet Chestnut', page 232).

Phosphorus requirements are not high, especially if healthy mycorrhizal fungi are present.

Any fertilizers used may well have some phosphorus in them. Wood ash is a good source, spread thinly beneath trees in spring and summer.

Walnuts are deep-rooting, and irrigation shouldn't be necessary unless rainfall is under 60cm (2') per year, or is particularly uneven in spread. Moisture shortage early in the season leads to small nuts; a deficit later can lead to a failure to mature wood. If irrigation is ever given, it must avoid wetting the foliage, as this will favour attacks of walnut blight – drip or trickle irrigation is most appropriate.

Pruning

The best time to prune walnuts to avoid sap 'bleeding' profusely from pruning cuts is midsummer to early winter. This sap leakage may be upsetting to the grower, but there's no evidence of it actually causing harm to walnut trees.

The aims of initial pruning in the first 5-7 years are:

* To prune off enough low branches as trees grow, to facilitate harvest beneath the trees. Also, if you are thinking ahead, for the trunk to eventually have veneer-making value, a minimum clean trunk of 2.1m (7') is needed.
* To develop a strong structure of 4-6 radiating main branches with large branch angles.

Ongoing pruning may be needed:

* For cultivars susceptible to walnut blight, to allow good aeration of trees.
* To allow for some light penetration so that productive fruiting wood is maintained in the lower part of the tree.

❋ To remove dead, diseased or crossing branches (very often the only pruning needed later on).

Walnut fruiting spurs live for 8-10 years, so fruit-wood renewal is less critical than with many fruit crops.

Pests

The main pests are the usual suspects – squirrels and birds (e.g. crows). Both can take nuts off the trees before fully ripe (they will eat through the strongly aromatic husks to get at them). See Chapter 2, page 53 for control measures.

Codling moth (*Laspeyresia pomonella*, syn. *Cydia pomonella*) is a major insect pest for commercial growers in warm climates, where it can destroy up to 30 per cent of the crop. Damage occurs in two ways: by early-season destruction of the kernel, or by late season shell and kernel staining as a result of feeding in the husk. Control may be aided by using pheromone traps to trap the male moths from mid-spring onwards; many commercial growers also spray insecticides. Attracting moth-consuming bats into the orchard can be hugely beneficial.

Diseases

The two main diseases of walnut are leaf blotch (anthracnose) and walnut blight. Honey fungus and blackline can also be problematical.

Leaf blotch or anthracnose (*Gnomonia leptostyla*)

Also known as common leaf spot fungus, this is present throughout Europe and North America.

The fungus causes brown blotches on leaves and sometimes young fruit; severe attacks result in defoliation and even blackening of the young green nuts, which fall prematurely. The disease appears in late spring and is favoured by wet weather. Blackened leaves fall from the tree (this is the easiest way of distinguishing this disease from walnut blight). Spores overwinter on dead leaves on the ground.

Raking up fallen leaves and burning or composting at high temperatures is a fairly good means of control. In wet seasons when infection is bad, copper-based sprays such as Bordeaux mixture are effective. However, the disease can be avoided by sensible cultivar choice.

Codling moth – a pest of walnut as well as apple.

Leaves affected by leaf botch (anthracnose).

Walnut blight (*Xanthomonas arboricola* pv. *juglandis*)

Also known as bacterial blight, this is a common and damaging disease wherever walnuts are grown on a large scale.

The bacterium causes small black angular spots, particularly towards the leaflet tips; large withered patches arise as they spread over the leaf surface (but leaves do not fall); black patches on shoots may lead to girdling and dieback; and black blotches and sunken lesions arise on the fruit. Up to 80 per cent of the crop may be lost in a bad attack, and the male catkins may also be destroyed.

The disease is most damaging when cool wet weather occurs at flowering time. Nuts may be attacked at any time during the season, but infection takes place almost exclusively during wet weather. Older wood is not subject to the disease, and even new shoots outgrow susceptibility in time.

The bacterium overwinters in healthy dormant buds and catkins, and can readily infect young shoots through wounds. Early-leafing cultivars are susceptible to earlier and more serious attacks of the disease, so sensible cultivar selection can largely avoid the problem. If the disease appears:

1. Ensure the pH is over 6.0 – attacks are worse on acid soils.
2. Avoid wetting foliage with irrigation.
3. Avoid excess nitrogen application.
4. Prune only lightly.

Honey fungus (*Armillaria* spp.)

Also known as shoestring fungus, oak root fungus and mushroom root rot, this widespread fungus is serious because it can attack almost anything, but it poses less of a threat to walnut than the two main diseases. Symptoms of attack are poor terminal growth, small yellowish leaves, premature defoliation, decreased productivity, shoot dieback and subsequent death.

The fungus causes root, collar and butt rot, and enters through roots or at the root collar. The disease can spread from an infected tree to a nearby healthy tree by means of leathery, bootlace-like rhizomorphs (root-like growths that some fungi produce). Apart from removing the diseased tree and roots, there are no control measures. Choice of rootstock can give some protection: northern California walnut (*J. hindsii*), black walnut (*J. nigra*) and hybrids are resistant, whereas walnut (*J. regia*) is susceptible. The risk must be weighed up against that of blackline (see below), for which grafting to *J. regia* rootstocks is the only control.

Blackline

Blackline disease occurs in grafted trees, when black lines and a delayed graft incompatibility form at the point of graft union. It is due to an

A walnut fruit infected with walnut blight.

273

infection by a strain of the cherry leaf roll virus (called CLRV-W), when rootstocks other than walnut (*J. regia*) are used, so is encountered most often in North America.

Harvesting and yields

Grafted cultivars usually start bearing female flowers and fruit in 2-5 years, though male flowering often commences a year or two later. Like other nuts, harvest is from the ground beneath trees – make sure grass or other understorey plants are cut short to facilitate walnut pick-up by hand tools or machine.

The husks around the nuts split at maturity in mid- to late autumn and allow the nuts to drop free. Shaking or tapping the branches may aid the drop of nuts – useful if windless conditions prevail for a while. Commercial operations often use a tree shaker to loosen most of the nuts in one or two passes.

The nuts can be collected by laying out heavy-duty nets, using Nut Wizards for hand-harvest or using mechanical harvesters commercially. Whichever way, nuts should be collected frequently and not allowed to lie on the ground for more than a day or two, as they are prone to both predation and fungal infection.

With natural nut fall, harvest for a specific cultivar is usually over a period of 10-14 days. Very little green husk remains attached to the nuts as they are harvested, and the nuts can be used, dried or sold fresh immediately. Commercial operations using tree shakers usually bring down most of the nuts, a proportion of which still have green husk attached, so the nuts have to be mechanically scrubbed to remove this.

Trees start cropping at about 5 years, and yields increase steadily until the trees are 25 years old or more (see table opposite). Crops are normally borne annually (there isn't a problem of biennial cropping), weather conditions permitting. In climates with cool, cloudy summers, young trees sometimes produce imperfect nuts – not sealed or with soft spots of hole in the shell – though the kernels are usually sound.

Walnut fruit for pickling are hand-picked from trees just after midsummer, when fruit have grown to full size but are still soft enough to push a needle through. On larger trees the fruit has to be cut off with a long pruning pole or knocked off, though the latter can cause some tree damage. Yields (because fruit are much heavier than the nuts) are around three times those above. Pickling companies making commercial pickled walnuts are keen to purchase the fruit if it can be transported to them quickly.

Husk splitting to release the walnut.

Walnut yields (assuming spacing of 12m/40')			
Age of tree	Yield per tree	Yield per hectare	Yield per acre
Young trees (5-10 years old)	2-5kg (4lb 6oz-11lb)	140-350kg	120-310lb
10-20 years	15-50kg (33-110lb)	1-3.5t	900-3,000lb
25 years plus at full crop	50-75kg (110-165lb)	3.5-5.2t	3,000-4,500lb

Processing and storage

Fresh walnuts grown in temperate climates (sometimes sold as 'wet walnuts') will not store for long – one or two weeks – before starting to mould on the outside. Of course, if grown in a climate with much drier, warmer autumn weather, they will have less water in the kernel and shell, and store longer.

Dry for long-term storage (see Chapter 3, page 70). Well-dried walnuts will store in-shell for 2-3 years at 15°C (59°F), longer at cooler temperatures.

Propagation

Seedling trees can be grown easily from seed. There are about 100 seeds/kg (45 seeds/lb); seed should be stratified (placed in cold, moist conditions) for 16 weeks before sowing. Nursery beds or sowing containers need very good protection from rodents, squirrels and birds. Seedlings of known parentage where both parents are good cultivars may inherit some or all the characteristics of their parents, very occasionally out performing their parents, and start bearing in 4-8 years.

Propagation of cultivars requires grafting. Like other *Juglans* species, walnuts are difficult in that they need significant heat at the graft union for grafts to heal. See Chapter 2, page 60 for suitable techniques.

Cultivars

There are several thousand walnut cultivars worldwide (with 800 or so from China alone!) and it is impossible to list them all here. The table below includes most of the commercial and home-garden cultivars from Europe, the Americas and Australasia.

The nearer you are to maritime conditions, the more important it is to choose cultivars with resistance to leaf blotch and walnut blight, as these are both worse in humid climates. Suitable cultivars usually come from those same conditions – cultivars of continental climate origin tend to have less resistance. Hence in maritime conditions, cultivars from France and the Netherlands tend to be better suited than those from Germany, Italy or central North America. In my own orchard in England, the best cultivars are 'Broadview' (British Columbia, Canada/Poland); 'Buccaneer' (Netherlands); 'Corne du Périgord', 'Fernette', 'Fernor', 'Franquette', 'Marbot', 'Mayette', 'Meylanaise', 'Parisienne' (all France); 'Hartley' (USA); and '139' (Germany).

Walnut fruits with a mild infection of walnut blight.

Late leafing in spring is another attribute that is useful in some regions and locations.

When choosing cultivars, good pollination is essential. A few cultivars are very self-fertile, otherwise you should plant a combination of cultivars that are protandrous (male flowers open before females) and protogynous (female flowers open first). Particularly good pollinators with male flowering occurring over a long period include 'Buccaneer', 'Chaberte', 'Cisco', 'Glady', 'Leib Mayette', 'Pedro', 'Ronde de Montignac', 'San Jose Mayette', 'Soleze' and 'Woodland'.

The hardiness of cultivars varies widely: some California varieties are hardy to zones 7-8 (H4-5), while others (often from mountainous areas of Europe – Carpathian varieties) are hardy to zone 4 (H7). Similarly, the amount of winter chilling needed varies, and ranges from 400 hours for the low-chill, warm-climate cultivars, to over 1,000 hours for the cold-climate cultivars; the average is 800 hours. Lack of winter chilling can result in delayed foliation, poor yields and branch dieback, particularly with late-leafing cultivars; so it is a mistake to plant cold-climate cultivars in warm regions.

Cultivars with red-skinned kernels have been derived from the natural form *rubra*. From the initial selections 'Red Danube' and 'Rubra', several other cultivars have been derived, including 'Aufhauser Baden', 'Buzsaki Pirosbelu', 'Jinboxiang 1', 'Kirschnuss', 'Petra's Red Medac', 'Purpurovy', 'Red Rief', 'Red Seal', 'Rote# 139', 'Rote Gubler', 'Rote Linzer', 'Rote moselwalnuss G.509', 'Rote moselwulfnuss' and 'Rubis' (Noyer Rouge). Red-skinned kernels are striking and in demand for specialized confectionery, etc. However, red-skinned cultivars are rarely ever as productive as other cultivars, and most are grown only as garden trees – few are cultivated commercially.

There are a few dwarf walnut cultivars that will only grow 2-4m (6'6"-13') high and wide. These include 'Dwarf Karlik 3' and '5', 'Fertilis' (also called 'Strauchwalnuss', bush walnut or *J. fertilis*) and 'Mini Multiflora Nr 14'. Dwarf trees will only yield proportionally to their size (a few kg/lb), but do make it possible to grow walnuts in small gardens. A number of other cultivars are low in vigour, meaning they will eventually grow to some 10m (33') high and 8m (27') wide.

Key to walnut cultivar table

Leafing: Leafing-out times vary over a period of around 30-40 days for different cultivars. The date of day 0 will vary from region to region (e.g. in England it is about 1 April):

VE = very early (day 0)
E = early (day 5-7)
EM = early–mid (day 10-14)
M = mid (day 15-21)

ML = mid–late (day 20-27)
L = late (day 25-29)
VL = very late (day 30-40)

Protn: Ticked if protandrous flowering (males earlier)
Protog: Ticked if protogynous flowering (females earlier)

Season: Ripening times vary over a period of 30-40 days for different cultivars, with each category covering 6-7 days. The date of day 0 will vary from region to region (e.g. in England it is about 25 September):

VE = very early
E = early
EM = early–mid
M = mid
ML = mid–late
L = late

SF: Ticked if cultivar is very self-fertile. ✓p means partially self-fertile (will still crop better with cross-pollination).

Chill: Indicates winter chilling required:
L = low
ML = mid to low
M = mid
H = high

Blight: Indicates susceptibility or resistance to walnut blight:
VS = very susceptible
S = susceptible
R = resistant

Blotch: Indicates susceptibility or resistance to leaf blotch (anthracnose):
VS = very susceptible
S = susceptible
R = resistant

Recommended walnut cultivars

Cultivar	Origin*	Leafing*	Protn*	Protog*	Season*	SF*	Chill*	Blight*	Blotch*	Comments
'Adams 10'	USA		✓					R		Medium-size nuts, easy to crack.
'Albi'	Poland								R	Tree of low vigour, precocious (early fruiting). Nuts medium size.
'Alsószentiváni 117'	Hungary	M	✓		EM		H			Tree vigorous, precocious, high yielding. Nuts large, thin-shelled, good flavour.
'Alsószentiváni kései'	Hungary	ML	✓		EM		H			Tree vigorous, precocious, part lateral bearing, high yielding. Nuts medium–large, thin-shelled, good flavour.
'Amigo'	USA	EM		✓	EM		M			
'Amphyon'	Netherlands	L			M	✓	H	S	R	Recent cultivar, precocious, with some lateral fruiting, and nuts borne in clusters of 2-4. A moderate producer. Nuts ridged, thin-shelled, pale kernels, very good flavour.
'Apollo'	Czech Republic			✓	M	✓	H			Tree very vigorous, spreading. Nuts large, medium-thick-shelled, good quality and flavour.
'Ashley' – see 'Payne'										
'Aufhauser Baden'	Germany									Tree of medium vigour. Red-skinned kernels within good-sized nuts.
'Axel'	Netherlands									Open-spreading tree. Nuts double-sized, low in oil.
'Backa'	Serbia	E	✓							Tree of moderate vigour, heavy cropper, lateral bearing. Nuts medium–large, conical.
'Bauer 2'	Canada				E		H			Tree vigorous, productive. Nut large, round, good quality.
'Bilecik'	Turkey			✓			M			Tree partly lateral bearing, nuts large.
'Bohumil'	Czech Republic		✓		M					Tree vigorous, upright. Nuts large, thin-shelled, very good quality.
'Bolle Jan'	Netherlands	VL				✓p				Upright tree, nuts round, thick-shelled, excellent flavour.
'Bonifác'	Hungary	ML		✓	ML		H			Tree of low vigour, precocious, lateral bearing, heavy cropping. Nuts medium size, thin-shelled, good flavour.

Recommended walnut cultivars

Cultivar	Origin*	Leafing*	Protn*	Protog*	Season*	SF*	Chill*	Blight*	Blotch*	Comments
'BP 09'	Montenegro	L			E				R	Tree of medium vigour, nuts large, good quality.
'Broadview'	British Columbia / Poland or Ukraine	ML	✓		M	✓p	H	S	R	Bears good crops at a very young age. Somewhat susceptible to walnut blight, but the heavy-cropping habit can still make it worth growing.
'Buccaneer'	Netherlands	L	✓	✓	E	✓	H	R	R	Tree vigorous, upright, bears large crops of round nuts. A good pollinator.
'Bucklov'	Czech Republic				E					Tree of medium vigour, very productive, frost-resistant. Nuts very large, medium-thick-shelled, good flavour.
'Bukovina 1'	Ukraine					✓	M		R	Tree moderately vigorous, high yielding, lateral bearing. Nuts medium size, thin-shelled.
'Bukovina 2'	Ukraine		✓				ML		R	Vigorous tree, high yielding. Nuts large, medium-thick-shelled.
'Bukovynska bomb'	Ukraine			✓			ML		R	Vigorous tree, medium yielding. Nuts large, thick-shelled.
'Burtner'	USA	VL				✓	H			
'Caesar'	USA			✓			H			
'Carmelo'	USA	L			M	✓	H	R		
'Chaberte'	France	VL	✓		L	✓p				
'Chambers'	USA	ML			M	✓	H			
'Champion of Ixworth'	UK	M								
'Chandler'	USA	ML	✓		M	✓p	M	R	R	Moderately vigorous tree, lateral bearing, heavy cropper. Nuts thin-shelled.
'Chernivtsi 1'	Ukraine			✓	M				R	Vigorous tree, high yielding, lateral bearing. Nuts medium size, round, thin-shelled.
'Chernivtsi 2'	Ukraine				ML					Vigorous tree, high yielding. Nuts large, thin-shelled.
'Chiara'	Netherlands				ML	✓			R	Tree precocious, nuts medium–large, good flavour.
'Chico'	USA	VE		✓	E		M	S		Tree small, upright, very productive, good-quality nuts.
'Cisco'	USA	L		✓	ML		M			Moderately vigorous tree, moderate yielding, often grown as a pollinizer.

Recommended walnut cultivars

Cultivar	Origin*	Leafing*	Protn*	Protog*	Season*	SF*	Chill*	Blight*	Blotch*	Comments
'Clarence'	USA			✓		✓p				
'Clinton'	USA			✓		✓p	H			
'Cobles #2'	USA	M	✓		M		H	S		Tree of moderate vigour, reasonably productive. Nuts double-sized, reasonably well filled. The pellicle on fresh kernels lacks any bitterness.
'Coenen'	Netherlands	E				✓p			R	Vigorous tree. Nuts medium size, good flavour.
'Colby'	USA	L		✓	E	✓	H			
'Combe'	USA	E					H	R		Very productive tree. Nuts medium size, oval, good quality.
'Corne du Périgord'	France	L		✓	ML		H	R	R	Old variety, still grown commercially, vigorous tree. Bears medium-size ridged nuts.
'Cyril'	Netherlands	ML			M	✓p			R	Vigorous upright tree, productive, nuts medium–large.
'Derning'						✓	H			
'Dionym'	Netherlands	L	✓	✓	E	✓	H	R	R	Recent selection. A very vigorous tree, precocious, some lateral fruiting, very productive. Nuts in clusters of 2-5, ridged, medium-thick-shelled, excellent flavour.
'Dodo'	Poland									Tree of medium vigour, precocious, high yielding, frost-resistant. Nuts medium size.
'Dooley 69-E'	USA			✓			H	R		Believed to be a hybrid with *J. nigra*, a healthy productive tree. Nuts medium size, thin-shelled, good flavour.
'Dublin's Glory'	New Zealand	E			E		H			Early flowering tree with good-quality nuts.
'Dwarf Karlik 3'	Ukraine			✓	E					Tree dwarf, growing only 2-3m (6'6"-10') high. Nuts medium size, excellent flavour.
'Dwarf Karlik 5'	Ukraine		✓		E					Tree dwarf, growing only 2-3m (6'6"-10') high. Nuts medium-sized, excellent flavour.
'Early Erhardt'		VE					L			
'Elit'	Slovenia	L					M		R	Tree of low vigour, productive, nuts medium size.
'Esterhazy II'	Hungary	ML	✓		M		M	S	S	Nuts medium size, good flavour. Susceptible to frost damage in spring.

Recommended walnut cultivars

Cultivar	Origin*	Leafing*	Protn*	Protog*	Season*	SF*	Chill*	Blight*	Blotch*	Comments
'Eureka'	USA	M			M	✓p	M			Vigorous tree, fair yields, an old cultivar with distinctive elongated nut shape.
'Europa'					L					Dwarf tree, very compact, precocious. Nuts medium size, good flavour.
'Excelsior of Taynton'	UK	EM	✓							
'Feradam'	France	E			E			S	S	Very recent cultivar ('Adams 10' x 'Chandler'). Tree of low vigour, moderately productive, lateral bearing. Nuts large, medium shell thickness, good quality.
'Ferbel'	France	E			EM					Very recent cultivar ('Chandler' x 'Lara'). Tree of medium vigour, moderately productive, lateral bearing. Nuts medium–large, excellent quality.
'Ferjean'	France	EM	✓				M	R	R	Fairly recent cultivar ('Grosvert' x 'Lara'). Lateral bearing, producing very heavy crops at a young age. Nuts medium size.
'Fernette'	France	L		✓	L		M	R	R	Fairly recent cultivar ('Franquette' x 'Lara'). Moderately vigorous, lateral bearing, good cropper, nuts round, large. A good pollinator.
'Fernor'	France	L	✓		L		M	R	R	Fairly recent cultivar ('Franquette' x 'Lara'). Lateral bearing, moderately vigorous, very good cropper of large nuts.
'Ferouette'	France	E			EM			R	R	Very recent cultivar ('Franquette' x 'Howard'). Tree vigorous, lateral bearing. Nuts large, average shell thickness, good quality.
'Fertignac'	France	VL				✓	M	R	R	Very recent cultivar ('Ronde de Montignac' x 'Chandler'). Tree moderately vigorous, a good pollinator, lateral bearing. Nuts medium–large, like larger 'Ronde de Montignac' nuts.
'Fertilis' (syn. 'Strauchwalnuss')	Germany									A dwarf tree growing to 5m (16') high and 3-4m (10-13') across; precocious. Nuts small, round, thin-shelled, good flavour.
'Fickes'			✓				H			

Recommended walnut cultivars

Cultivar	Origin*	Leafing*	Protn*	Protog*	Season*	SF*	Chill*	Blight*	Blotch*	Comments
'Forde'	USA	M		✓	ML		M	R		Tree moderately vigorous, lateral bearing, good cropper, nuts with strong shells. Very recent Californian introduction.
'Franquette'	France	VL	✓		L	✓p	M	R	R	Old variety, grown commercially. Vigorous tree, bearing quite good crops of medium–large nuts. Has some resistance to codling moth.
'Geisenheimer' (syn. 'Nr. 26')	Germany	VL	✓		M	✓	H		VS	Tree of low vigour, nuts medium–large. Exhibits apomixis (asexual seed formation), enabling nuts to develop without pollination.
'Gillet'	USA	M		✓	EM		M	R		A moderately vigorous tree, lateral bearing, good cropper. Large nuts of good quality. Very recent Californian introduction.
'Gizavezhda'	Afghanistan								S	Late-flowering tree resistant to codling moth with medium-size nuts.
'Glady'	France	VL		✓	L					
'Grandejean'	France	L			EM					Vigorous tree, slow to bear, nuts small but well flavoured.
'Gultekin 1'	Turkey						M			Very large nuts.
'Gustine'		E			E			S		
'Hansen'	Germany	EM	✓		E	✓	H	R		Produces medium-size nuts. Tree small.
'Hartley'	USA	ML	✓		EM	✓p	M	R	R	Vigorous tree, good cropper. Nuts with distinctive conical/triangular shape. Susceptible to deep bark canker. One of the best Californian cultivars to grow elsewhere.
'Helmle'	USA	VL					H			
'Holton'	USA	EM				✓	H			
'Howard'	USA	ML	✓		M	✓p	M	S		Moderately vigorous tree, lateral bearing, heavy cropper with large nuts.
'Ibar'	Serbia	L								Tree of moderate vigour and production, nuts medium size.
'ISU 73-H-24'	USA			✓			H	R		Productive tree, nuts large, oval, thin-shelled, good quality.

Recommended walnut cultivars

Cultivar	Origin*	Leafing*	Protn*	Protog*	Season*	SF*	Chill*	Blight*	Blotch*	Comments
'Ivanhoe'	USA	E		✓	EM		M	S		Smaller tree of low vigour, lateral bearing, very precocious and yielding well. Nuts thin-shelled.
'Izvor 10'	Bulgaria							R	S	Tree of moderate vigour, lateral bearing, nuts medium–large.
'Jacek'	Poland									Vigorous tree, frost-resistant, very large nuts (double-sized).
'Jacobs'		L								
'Jupiter'	Czech Republic	L	✓		ML		H			Vigorous tree, frost-resistant. Medium–large nuts, medium-thick-shelled, good quality and flavour.
'Kaman 1'	Turkey			✓			M			Tree lateral bearing, medium-size nuts.
'Kaplan 86'	Turkey						M			Very large nuts (double-sized).
'Kasni Grozdasti'	Serbia	VL	✓							Tree of low vigour, nuts medium size.
'Kasni Rodni'	Serbia	L	✓		E	✓			R	Tree of low vigour, medium–large nuts.
'Keener'		L								
'Kentucky Giant'	USA	L			E		H			
'Klishhovsky'			✓		E				R	
'Klosz'	Switzerland									Nuts large, thin-shelled, very good quality.
'Koszycki'	Poland									Old cultivar often propagated by seed. Tree of low vigour with very large double-sized nuts.
'Kurmarker' (syn. 'Nr. 1247')	Germany	EM		✓	EM		H	R	VS	Vigorous tree with large nuts of excellent flavour.
'Lady Irene'	UK	EM	✓				H	VS		
'Lake'	USA	L		✓	M	✓	H	R		Productive tree widely planted in eastern North America, the Czech Republic and Slovakia. Tree vigorous, spreading, frost-resistant. Nuts medium size, medium-thick-shelled, very good quality.
'Lancaster'			✓				H			
'Lara'	France	EM	✓				H	S		Recent variety. Low vigour, lateral bearing, precocious, heavy cropper. Large, sweet nuts.
'Leib Mayette'	USA	VL	✓		ML		H			

Recommended walnut cultivars

Cultivar	Origin*	Leafing*	Protn*	Protog*	Season*	SF*	Chill*	Blight*	Blotch*	Comments
'Leopold'	Poland									Tree of medium vigour with serrated leaves, frost-resistant. Nuts medium size, very good flavour.
'Leshnica'	Afghanistan	E			M					Vigorous tree, lateral bearing, nuts large, thin shelled, good quality.
'Littlepage'			✓			✓p	H			
'Lompoc'	France	E			E					
'Magdon'					M					
'Manregian'	USA	M					H	R	R	
'Maras 18'	Turkey						M			Tree lateral bearing, large nuts.
'Marbot'	France	L		✓	ML		H	R	R	Old variety. Tree of moderate vigour, good cropper, bears medium to large nuts.
'Mars'	Czech Republic	L			M	✓	H		R	Vigorous tree, frost-resistant, nuts large, thin-shelled, very good quality, sweet flavour.
'Mayette'	France	VL	✓		M		M	R	R	Old variety. Vigorous tree, bears medium–large nuts.
'McDermid'	USA	E	✓			✓p	H			
'McKinster'	USA	M		✓		✓p	H			
'Metcalfe'	USA	L	✓			✓	H			
'Meylanaise'	France	VL		✓	ML	✓	H	R	R	Old variety. Bears good crops of medium-size, round nuts. A good pollinator.
'Meyrick'	New Zealand	L						R		Tree of moderate vigour. Nuts large, thin-shelled, good quality.
'Midland'		M			EM					
'Milotai 10'	Hungary	EM	✓		EM		H			Tree vigorous, part lateral bearing, productive. Nuts large, thin-shelled, good flavour.
'Milotai bőtermő'	Hungary	ML	✓		M		H		S	Tree vigorous, precocious, productive. Nuts large, thin-shelled, good flavour.
'Milotai intenzív'	Hungary	EM		✓	ML		H	VS	S	Tree of low vigour, precocious, lateral bearing, heavy cropping. Nuts medium size, thin-shelled, good flavour.
'Milotai kései'	Hungary	VL		✓	ML		H	S	R	Tree moderately vigorous, precocious, lateral bearing, heavy cropping. Nuts medium size, thin-shelled, good flavour.

Recommended walnut cultivars

Cultivar	Origin*	Leafing*	Protn*	Protog*	Season*	SF*	Chill*	Blight*	Blotch*	Comments
'Mini Multiflora Nr. 14'	Netherlands					✓				Tree a dwarf, precocious, nuts medium size, good flavour.
'Mire'	Serbia	E	✓				M			Tree of moderate vigour, very heavy cropper. Nuts medium to large, very good quality.
'Morris'				✓		✓p	H			
'Moselaner' (syn. 'Nr. 120')	Germany	ML	✓		M		H		VS	Vigorous tree. Nuts medium–large.
'Newgates Mayette'	USA	VL		✓	ML		M			
'North Platte'	USA			✓			H	R		Vigorous tree, good cropper. Nuts medium-large, good quality.
'Northdown Clawnut'	UK	EM	✓							
'Nr. 16'	Germany		✓		**ML**	✓	H	R	**VS**	Tree of medium vigour. Nuts medium size, pointed, good flavour.
'Nr. 26' – see 'Geisenheimer'										
'Nr. 120' – see 'Moselaner'										
'Nr. 139' – see 'Weinheimer'										
'Nr. 286' – see 'Spreewalder'										
'Nr. 1239' - see 'Red Danube'										
'Nr. 1247' – see 'Kurmarker'										
'Oguzlar 77'	Turkey						M			Tree lateral bearing, medium-size nuts.
'Orth'			✓			✓p	H			
'Ovcar'	Serbia									High-yielding tree with large nuts.
'Parisienne'	France	L	✓		EM		H	R	R	Old variety. Vigorous tree, bears medium–large conical nuts, good in cool climates.
'Patching'	USA	EM	✓		E					
'Payne' (syn. 'Ashley')	USA	E	✓		EM	✓p	M	VS		Tree of moderate vigour, upright, precocious, lateral bearing, moderate yielding. Nuts of good quality.

Recommended walnut cultivars

Cultivar	Origin*	Leafing*	Protn*	Protog*	Season*	SF*	Chill*	Blight*	Blotch*	Comments
'Pedro'	USA	M	✓		ML	✓	L	VS		Tree of moderate vigour, nuts of good quality but kernels susceptible to darkening after harvest.
'Placentia'		E			M		L			
'Plovdivski'	Bulgaria		✓		E		H	R	R	Fairly vigorous tree, precocious. A good cropper, nuts of very good quality, large kernels.
'Proslavski'	Bulgaria		✓		E		H	R	R	Vigorous tree with large leaves. Bears good crops of large nuts with large kernels.
'Ramilette'	Argentina									Tree highly productive, lateral bearing with large nuts.
'Rasna'	Serbia	L			E	✓	M			Tree of low to medium vigour, very productive. Nuts large.
'Red Danube' (syns. 'Nr. 1239', 'Rode Donau')	Germany	ML		✓	M		H		VS	Tree of moderate vigour, lateral bearing, moderate yielding. Nuts small–medium, kernels with red skin.
'Red Gubler I – IV'	Switzerland									A series of cultivars (I – IV) with red-skinned kernels.
'Resovia'	Poland									
'Rex'	New Zealand	L								Tree of medium vigour, frost-resistant, high yielding. Nuts medium size.
'Rita'	USA/Poland			✓			H	R		Vigorous tree, very productive, nuts small and good quality.
'Robert Livermore'	USA	ML	✓		M		M			Of Carpathian origin, tree small, bears heavy crops of thin-shelled nuts. Very hardy.
'Rode Donau' – see 'Red Danube'										
'Ronde de Montignac'	France	VL	✓	✓	M	✓	H	R	R	
'Rudkovsky'	Ukraine				M					An old variety, fairly vigorous, bears medium-size nuts. A good pollinator.
'Sampion'	Serbia	EM	✓				H			Tree vigorous, high yielding, disease-tolerant. Nuts medium–large, thin-shelled.
'San Jose Mayette'	USA	L	✓		M	✓p	H	R		Tree of moderate vigour, very productive, extremely hardy, nuts large.

Recommended walnut cultivars

Cultivar	Origin*	Leafing*	Protn*	Protog*	Season*	SF*	Chill*	Blight*	Blotch*	Comments
'Saturn'	Czech Republic		✓		ML	✓	H			Recent Czech variety. Tree moderately vigorous, good cropper, nuts medium–large, medium-thick-shelled, good quality, sweet flavour.
'Sava'	Serbia	M	✓							Tree of low vigour. Nuts large.
'Schafer'		L			VE		H			
'Scharsch Franquette'	USA	VL	✓		ML	✓p	M			
'Schwartz Franquette'	USA	VL	✓		ML	✓p	M			
'Sebin'	Turkey		✓				M			Tree high yielding, medium-size nuts.
'Seifersdorfsky'	Czech Republic	EM			M	✓p	H		R	Tree very vigorous, spreading. Nuts large, medium-thick-shelled, very good quality and flavour.
'Sejnovo'	Serbia	L	✓		M		H	R		Tree moderately vigorous and productive. Nuts large, oval, thin-shelled, good quality, high oil content.
'Sen 1'	Turkey					✓	M			High-yielding tree, large nuts.
'Serr'	USA	EM		✓	EM	✓p	M	S		Very vigorous tree, lateral bearing, good yields, nuts thin-shelled. Susceptible to flower drop where there is excessive pollen.
'Sexton'	USA	EM	✓		M		M	R		Densely branched tree, lateral bearing with clusters of flowers, very precocious and heavy yielding. Nuts large and round. Very recent Californian introduction.
'Sibisel 8'	Romania			✓						Productive tree, nuts medium size.
'Silesia'	Poland									Very vigorous tree, precocious, nuts medium size.
'Smokthina'	Afghanistan	M			M			S		Vigorous rounded tree, medium yields, lateral bearing. Nuts medium size, good quality.
'Solano'	USA	M		✓	EM		M			Tree moderately vigorous, lateral bearing, heavy cropping. Good-quality nuts.

Recommended walnut cultivars

Cultivar	Origin*	Leafing*	Protn*	Protog*	Season*	SF*	Chill*	Blight*	Blotch*	Comments
'Soleze'	France	L	✓	✓	M	✓p	H	R	R	An old variety. A large tree, heavy cropping. A good pollinator.
'Spreewalder' (syn. 'Nr. 286')	Germany	EM	✓		E		H		VS	Tree of medium to high vigour, nuts very large.
'Spurgeon'	USA	VL	✓		L	✓p	H			
'Srem'	Serbia	E	✓							Tree of moderate vigour, heavy cropper. Nuts very large, thin-shelled.
'Stan'	New Zealand	L								Moderately vigorous tree, not precocious, nuts elongated, small–medium, very good flavour.
'Strauchwalnuss' – see 'Fertilis'										
'Stutton Seedling'	UK		✓					VS		
'Sunland'	USA	E		✓	ML		M	VS		Vigorous tree, lateral bearing, good cropper of large, oval nuts.
'Sutyemez 1'	Turkey			✓			M			Tree lateral bearing, very large nuts (double-sized).
'Sychrov'	Czech Republic			✓	EM		H			Tree vigorous, spreading. Nuts medium size, medium-thick-shelled, quite good quality.
'Targo'	Poland									Tree of low to medium vigour, precocious, frost-resistant. Nuts medium to large.
'Tehama'	USA	M	✓		M	✓p	M	R		Vigorous tree, lateral bearing, heavy cropping.
'Tisa'	Serbia	E				✓	M			Tree of moderate vigour, crops well. Nuts borne in clusters of 7-8, large, round, thin-shelled.
'Tiszacsécsi 83'	Hungary	L	✓		M		H			Tree vigorous, open, partly lateral bearing, precocious, heavy cropper. Nuts medium size, thin-shelled, good flavour.
'Treyve'	France	VL	✓							
'Trinte'		E			EM					
'Trompito'	Argentina	E								Moderately vigorous tree, heavy cropper, lateral bearing. Nuts medium size.
'Tryumf'	Poland								R	Vigorous tree, precocious, frost-resistant. Medium-size nuts.
'Tulare'	USA	M	✓		M		M			Vigorous upright tree, lateral bearing, heavy cropping. Good-quality nuts.

Recommended walnut cultivars

Cultivar	Origin*	Leafing*	Protn*	Protog*	Season*	SF*	Chill*	Blight*	Blotch*	Comments
'Victoria'	Czech Republic			✓	M		H			Tree vigorous. Nuts large to very large, thin-shelled, very good quality.
'Vilem'	Czech Republic		✓		L		H			Tree moderately vigorous. Nuts medium–large, medium-thick-shelled, very good quality.
'Vina'	USA	EM	✓		EM		M	R		Tree moderately vigorous, lateral bearing, heavy cropper. Nuts good quality, conical.
'Vourev'		VL			L					
'Waterloo'		L					M			
'Weinheimer' (syn. 'Nr. 139')	Germany	M	✓		E		H	R	S	Tree moderately vigorous, frost-resistant. Nuts medium size. Exhibits apomixis – enables nuts to develop without pollination.
'Weinsberg 1' – see 'Weinsburger'										
'Weinsberg 2'	Germany				E			R		Tree of moderate vigour. Nuts large, thin-shelled.
'Weinsburger' (syn. 'Weinsberg 1')	Germany	VE	✓		EM		M			Tree of low vigour. Large nuts, thin-shelled, good flavour.
'Wilsons Wonder'	New Zealand							S		Tree with very large nuts.
'Woodland'	USA			✓						
'Wunder von Monrepos'	Germany	L	✓			✓			R	Tree of moderate vigour, productive. Excellent-quality nuts.
'Yablunovsky'	Ukraine				E				R	Tree of low vigour, high yielding, precocious. Nuts medium size, thin-shelled.
'Yalova 1'	Turkey		✓				L			Very large nuts.
'Yalova 3'	Turkey		✓				L			Medium–large nuts.
'Yalova 4'	Turkey					✓	M			Medium–large nuts.
'Yarivsky'	Ukraine			✓			ML		R	Tree of low vigour, high yielding. Nuts large, thin-shelled.
'Youngs B1'	USA			✓			H			Tree vigorous, productive. Nuts large, round, good quality. A 'Broadview' seedling.

Where no details are given, this is because information is not readily available for these cultivars.

YELLOWHORN

(*Xanthoceras sorbifolium*)

ZONE 4, H7

Also known as shiny-leaf yellowhorn and northern macadamia, yellowhorn is a deciduous large shrub or small tree that is native to northern China, where it grows in thickets on dry hill slopes. The name refers to the horn-like growths between the petals.

Yellowhorn coming into leaf in spring.

In the wild, yellowhorn grows up to 6m (20') high and 2.5m (8') wide, upright at first but becoming rounded with age; in cultivation in cool climates, it rarely exceeds 3m (10') high. Yellowhorn is very tough once established, and can live for 100-200 years.

The leaves are alternate, up to 30cm (1') long with 9-17 deeply toothed leaflets, each 4-5cm (1⅝-2") long, dark green above and paler green beneath. They persist into the autumn, falling very late.

Dense clusters of white-petalled flowers, 2-3cm (¾-1⅛") across, are produced on flowering shoots up to 25cm (10") long in mid-spring, at the same time as the leaves emerge. They are flushed yellow, later carmine at the base, and are sweetly scented.

Fruit are leathery, thick-walled oval capsules, 4-6cm (1⅝-2½") across, which enclose 8-10 round brown nuts. The capsules eventually split to release the nuts, which are 10-15mm (⅜-⅝") across and look like small, dark brown hazelnuts.

Uses

The seeds are edible, with a brazil nut flavour and crunchy texture. They can be eaten raw, roasted and shelled, or cooked in a wide range of dishes.

An edible oil can be pressed from the nuts.

Secondary uses of yellowhorn

In China yellownut oil is being investigated as a prospective component in biodiesel.

The inner bark and thick fleshy roots are yellow and can probably be used for dyeing.

Cultivation

Yellowhorn prefers a continental climate with cold winters and hot summers. It will tolerate more temperate climates but is slow-growing, especially when young.

The tree requires a long, warm growing season to fully ripen its wood and stimulate the production of flower buds. Give protection from cold winds; late spring frosts can be damaging. In cool regions, a protective wall may be needed for flowering to occur.

Plants are usually slow to become established and grow to about 2.4m (8') high in 10 years. Plant at about 3m (10') apart.

Rootstocks and soils

Yelllowhorn likes a well-drained fertile soil and a sunny position, though most soils (including chalk) apart from wet ones are tolerated.

Pollination

Flowers are produced on the previous year's wood in late spring. Pollination is via bees and other insects. Plants are self-fertile.

Yellowhorn flowers.

Feeding and irrigation

Plants that are cropping well should be fed with a general-purpose fertilizer, compost or manure. This will encourage the growth of spurs and the development of flowers / young fruit the next year.

In dry summer regions, irrigation may be necessary to sustain good cropping.

Pruning

None needed, except cutting out any diseased wood.

Pests

None of note.

Diseases

Susceptible to attacks by coral spot fungus (*Nectria cinnabarina*), particularly if the wood is not fully ripened. Affected wood should be cut out.

Harvesting and yields

Flowering and fruiting begins at about five years of age. In the autumn, pick whole fruits off the bushes as they start to split; it is then easy to release the nuts.

The fruit opens to release the nuts.

Yellowhorn nuts.

Processing and storage

Dry nuts for long-term storage of a few years (see Chapter 3, page 70). Shells are leathery (like sweet chestnut) but become brittle once dried.

Propagation

The seeds require 6 weeks of cold stratification before sowing in warm temperatures. Most germination occurs within 4-6 weeks but some seeds can take several months.

Root cuttings can be taken in midwinter. Cut pieces 3cm (1⅛") long and plant them horizontally in pots under cover. A good percentage usually take.

Division of suckers is possible in winter.

Cultivars

There are none available outside China.

Glossary

Abscission: The shedding of a nut or other part of a plant.

Agroforestry: The growing of trees with other crops and/or animals beneath.

Air-pruning: A propagation technique used to stimulate a vigorous root system. Young plants are grown in open-bottomed containers, so that root tips are exposed to dry air and die back. In response the plant sends out an abundance of secondary roots, which then spread through the container.

Allelopathic: Where one plant produces biochemicals that influence (usually negatively) the survival and reproduction of other organisms.

Alley cropping: Growing lines of trees, with alleys between to grow annual crops such as cereals or vegetables.

Apomixis: The ability to produce seeds/nuts without pollination.

Bearing surface: The surface area of a plant that will bear flowers and fruits/nuts.

Biennial: Flowering and fruiting every second year.

Burr: A rough or spiny covering around a nut.

Chilling requirement: Most trees grown outside the tropics and subtropics have a chilling requirement – that is, in the winter they need a minimum number of hours with temperatures below 7°C (45°F). Without enough chilling, these trees flower and fruit erratically.

Coppicing: A cyclic management technique of cutting young trees down near to the ground and allowing regrowth of strong new shoots.

Dioecious: Where male and female flowers are found on different individual plants (usually called male or female plants or cultivars).

Etiolated: Shoots formed by a plant in the dark.

Feeder root: Small hair-like roots, which are very efficient at taking up water and nutrients from soil.

Frost pocket: An area where cold air travelling downhill is trapped, e.g. by a hedge or building.

Harden off: The process of new season growth becoming mature enough to survive the following winter.

Husk: A smooth leathery covering around a nut.

Intercropping: Growing a crop of one species among plants of a different kind.

Juglone: An allelopathic chemical found in the roots, leaves and green husks of many members of the walnut family (*Juglans* spp.). See *Allelopathic*.

Kernel: The internal edible part of a nut.

Layering: A method of propagation where a shoot is fastened down to form roots while still attached to the parent plant.

Mast year: A year in which the fruit/nut crop produced by a tree is significantly higher than usual.

Monoecious: Where male and female flowers are found on the same individual plant. This does not mean a plant is self-fertile.

Mycorrhizal fungi: Symbiotic fungi that associate with most trees, including nut trees.

Nitrogen-fixing plants: Plants that form a symbiotic relationship with bacteria (normally found in structures called nodules attached to the roots), allowing utilization of atmospheric nitrogen.

Panicle: A loose branching cluster of flowers.

Pellicle: The thin inner skin around the nut kernel, usually brown and bitter when fresh.

Perennial: A plant that lives for three years or longer. Usually refers to low herbaceous or evergreen plants rather than trees or shrubs.

Pollarding: A cyclic management technique of cutting young trees down to a height of around 1-2m (3-6'6") and allowing regrowth of strong new shoots.

Pollen incompatibility group: A group of cultivars that will not pollinate each other, usually because of genetic closeness.

Precocious: Flowering early in life.

Protandrous: Where male flowers mature before the female flowers.

Protogynous: Where female flowers mature before the male flowers.

Rootstock: The lower part of a grafted plant, growing in the ground and connected to a genetically different upper part. See *Scion*.

Sacrificial tree: A tree planted to lure a pest away from a more valuable crop.

Scion: The upper part of a grafted plant, having no roots of its own but connected to a genetically different lower part. See *Rootstock*.

Silvoarable: See *Alley cropping*.

Silvopasture: Growing widely spaced trees, with pasture or forage below for grazing animals.

Soil probe: An instrument to insert into soil to measure certain characteristics.

Sucker: Where a plant produces a new shoot from the ground, sometimes some distance away from the mother plant.

Tannin: A yellowish or brownish astringent organic substance present in some barks and other plant tissues.

Taproot: A straight tapering root, growing vertically downwards and forming the centre from which subsidiary rootlets branch.

Tensiometer: An instrument to measure soil moisture tension and thus soil moisture content and requirement for crops.

Tip-bearer: A tree or shrub bearing most or all of its flowers (and thus fruit/nuts) on very young growth near the tips of branches.

Top-work: Grafting on to the upper part of an already established tree to change the cultivar of one or more branches.

Undercutting: A propagation technique used in tree nurseries. A horizontal blade is drawn through the soil at a certain depth below a bed of young bare-rooted trees to sever the deeper roots. This encourages a more fibrous root system to develop and improves the quality of the trees when planted out. Sometimes this can also be achieved by using a spade at an angle.

Wands: Long, flexible year-old shoots of hazel or other species.

Appendix 1: Nutritional content of nuts

The following tables give a detailed breakdown of the nutritional content of starchy nuts and oily nuts.

Average nutritional content of nut kernels – starchy nuts (per 100g portion)

		Acorns (fresh)*	Acorns (dried)*	Acorns (range of 18 species)	Acorns (Q. robur)	Chinkapin	Ginkgo (fresh)	Ginkgo (dried)	Monkey puzzle	Sweet chestnut (C. crenata) raw	Sweet chestnut (C. crenata) dried	Sweet chestnut (C. mollissima) raw	Sweet chestnut (C. mollissima) dried	Sweet chestnut (C. sativa) raw	Sweet chestnut (C. sativa) dried
Water	g	27.9	5.06	8.7-44.6	13.9		55.2	12.4		61.41	9.96	43.95	8.9	52	9
Energy	kcal	387	509	265-577	353	478	182	348		154	360	224	363	196	369
Protein	g	6.15	8.1	2.3-8.6	7.9	~5	4.32	10.35		2.25	5.25	4.2	6.82	1.63	5.01
Total lipid (fat)	g	23.86	31.41	1.1-31.4	4.6	~5	1.68	2		0.53	1.24	1.11	1.81	1.25	3.91
Fatty acids, total saturated	g	3.102	4.084				0.319	0.381		0.078	0.183	0.164	0.266	0.235	0.736
Fatty acids, total monounsaturated	g	15.109	19.896				0.619	0.739		0.278	0.65	0.581	0.945	0.43	1.349
Fatty acids, total polyunsaturated	g	4.596	6.052				0.618	0.737		0.138	0.322	0.288	0.468	0.493	1.546
Carbohydrate	g	40.75	53.66	32.7-89.7	67.8	~40	37.6	72.45	64.0	34.91	81.43	49.07	79.76	44.17	78.43
Minerals															
Calcium, Ca	mg	41	54				2	20		31	72	18	29	19	64
Iron, Fe	mg	0.79	1.04				1	1.6		1.45	3.38	1.41	2.29	0.94	2.39
Magnesium, Mg	mg	62	82				27	53		49	115	84	137	30	74
Phosphorus, P	mg	79	103				124	269		72	169	96	155	38	137
Potassium, K	mg	539	709				510	998		329	768	447	726	484	991
Sodium, Na	mg	0	0				7	13		14	34	3	5	2	37
Zinc, Zn	mg	0.51	0.67				0.34	0.67		1.1	2.57	0.87	1.41	0.49	0.35
Vitamins															
Vitamin C, total ascorbic acid	mg	0	0				15	29.3		26.3	61.3	36	58.5	40.2	15.1
Thiamin	mg	0.112	0.149				0.22	0.43		0.344	0.802	0.16	0.26	0.144	0.354
Riboflavin	mg	0.118	0.154				0.09	0.176		0.163	0.38	0.18	0.293	0.016	0.054
Niacin	mg	1.827	2.406				6	11.732		1.5	3.5	0.8	1.3	1.102	0.854
Vitamin B6	mg	0.528	0.695				0.328	0.641		0.283	0.659	0.41	0.666	0.352	0.666
Folate, DFE	µg	87	115				54	106		47	109	68	110	58	110
Vitamin A, RAE	µg	2	0				28	55		2	4	10	16	1	0
Vitamin A, IU	IU	39	0				558	1,091		37	86	202	328	26	0

* It is unclear which species of oak these refer to. There is quite a lot of variability between acorns of different species, as can be seen from the next columns. There is also great variability in tannin levels in acorns from different species, ranging from 0.1g to 8.8g per 100g.

None of the starchy nuts contains Vitamin B12, Vitamin D / D2 / D3 or cholesterol.
The nutritional content of golden chinkapins is very similar to that of acorns (from oaks).

Average nutritional content of nut kernels – oily nuts (per 100g portion)

		Almonds	Black walnuts	Butternuts	Hazelnuts	Hickory nuts	Pecans	Pine nut (P. cembroides)	Pine nut (P. pinea)	Walnuts
Water	g	4.41	4.56	3.34	5.31	2.65	3.52	5.9	2.28	4.07
Energy	kcal	579	619	612	628	657	691	629	673	654
Protein	g	21.15	24.06	24.9	14.95	12.72	9.17	11.57	13.69	15.23
Total lipid (fat)	g	49.93	59.33	56.98	60.75	64.37	71.97	60.98	68.37	65.21
Fatty acids, total saturated	g	3.802	3.483	1.306	4.464	7.038	6.18	9.377	4.899	6.126
Fatty acids, total monounsaturated	g	31.551	15.442	10.425	45.652	32.611	40.801	22.942	18.764	8.933
Fatty acids, total polyunsaturated	g	12.329	36.437	42.741	7.92	21.886	21.614	25.668	34.071	47.174
Carbohydrate, by difference	g	21.55	9.58	12.05	16.7	18.25	13.86	19.3	13.08	13.71
Fibre, total dietary	g	12.5	6.8	4.7	9.7	6.4	9.6	10.7	3.7	6.7
Sugars, total	g	4.35	1.1		4.34		3.97		3.59	2.61
Minerals										
Calcium, Ca	mg	269	61	53	114	61	70	8	16	98
Iron, Fe	mg	3.71	3.12	4.02	4.7	2.12	2.53	3.06	5.53	2.91
Magnesium, Mg	mg	270	201	237	163	173	121	234	251	158
Phosphorus, P	mg	481	513	446	290	336	277	35	575	346
Potassium, K	mg	733	523	421	680	436	410	628	597	441
Sodium, Na	mg	1	2	1	0	1	0	72	2	2
Zinc, Zn	mg	3.12	3.37	3.13	2.45	4.31	4.53	4.28	6.45	3.09
Vitamins										
Vitamin C, total ascorbic acid	mg	0	1.7	3.2	6.3	2	1.1	2	0.8	1.3
Thiamin	mg	0.205	0.057	0.383	0.643	0.867	0.66	1.243	0.364	0.341
Riboflavin	mg	1.138	0.13	0.148	0.113	0.131	0.13	0.223	0.227	0.15
Niacin	mg	3.618	0.47	1.045	1.8	0.907	1.167	4.37	4.387	1.125
Vitamin B6	mg	0.137	0.583	0.56	0.563	0.192	0.21	0.111	0.094	0.537
Folate, DFE	µg	44	31	66	113	40	22	58	34	98

Average nutritional content of nut kernels – oily nuts (per 100g portion)

		Almonds	Black walnuts	Butternuts	Hazelnuts	Hickory nuts	Pecans	Pine nut (*P. cembroides*)	Pine nut (*P. pinea*)	Walnuts
Vitamin A, RAE	µg	0	2	6	1	7	3	1	1	1
Vitamin A, IU	IU	2	40	124	20	131	56	29	29	20
Vitamin E (alpha-tocopherol)	mg	25.63	2.08		15.03		1.4		9.33	0.7
Vitamin K (phylloquinone)	µg	0	2.7		14.2		3.5		53.9	2.7
Fatty acids, total saturated	g	3.802	3.483	1.306	4.464	7.038	6.18	9.377	4.899	6.126
Fatty acids, total monounsaturated	g	31.551	15.442	10.425	45.652	32.611	40.801	22.942	18.764	8.933
Fatty acids, total polyunsaturated	g	12.329	36.437	42.741	7.92	21.886	21.614	25.668	34.071	47.174

These values are for nuts harvested in dry weather, so fresh and dry nutritional content are more or less the same (unlike the starchy nuts above). After a wet harvest, nuts are likely to contain more water than is indicated here, but such a crop will usually be dried, in any case, to prevent moulds.

None of the oily nuts contain Vitamin B12, Vitamin D/D2/D3 or cholesterol.

The nutritional content of buartnuts is very similar to that of butternuts.
The nutritional content of heartnuts is very similar to that of walnuts.
The nutritional content of trazels is very similar to that of hazelnuts.

Sources of nutritional information

Crawford, M. (2016). Edible acorns from oaks. Factsheet N04, Agroforestry Research Trust.

Payne, J. et al (1994). 'Castanea pumila (L.) Mill.: An underused native nut tree'. *Hort. Science*, 29(2).

United States Department of Agriculture (USDA) food nutrient database: http://ndb.nal.usda.gov/ndb/foods

Appendix 2: Common and Latin names

The following table gives the common and Latin names for all the plants mentioned in this book. Those species with no common name (e.g. *Q. brantii*, *Q. ehrenbergii*) are not included in this table.

Commonly known as	Latin name	Also known as
Afares oak	*Quercus afares*	
Alder	*Alnus glutinosa*	Black alder, common alder
Aleppo oak	*Quercus infectoria*	
Aleppo pine	*Pinus halepensis*	
Almond	*Prunus dulcis*	
American bladdernut	*Staphylea trifolia*	
American chestnut	*Castanea dentata*	
American hazel	*Corylus Americana*	
American turkey oak	*Quercus laevis*	Scrub oak
Arizona white oak	*Quercus arizonica*	
Arolla pine	*Pinus cembra*	Swiss stone pine
Autumn olive	*Elaeagnus umbellata*	
Ballota oak	*Quercus ilex* var. *ballota*	Barbary oak
Bartram's oak	*Quercus x heterophylla*	
Bear oak	*Quercus ilicifolia*	Scrub oak
Bebb's oak	*Quercus x bebbiana*	
Big-cone pinyon	*Pinus maximartinezii*	
Bitternut hickory	*Carya cordiformis*	
Black cohosh	*Cimicifuga racemosa*	
Black locust	*Robinia pseudoacacia*	
Black oak	*Quercus velutina*	Smooth-bark oak
Black walnut	*Juglans nigra*	Virginian walnut, American walnut, eastern black walnut
Blackjack oak	*Quercus marylandica*	Jack oak
Bladdernut	*Staphylea pinnata*	European bladdernut, common bladdernut
Blue Japanese oak	*Quercus glauca*	
Blue oak	*Quercus douglasii*	Iron oak
Boz pirnal oak	*Quercus aucheri*	
Broom	*Cytisus scoparius*	Common broom
Buartnut	*Juglans x bixbyi*	
Bur English oak	*Quercus macroocarpa x robur*	
Bur Gambel oak	*Quercus macrocarpa x gambelii*	
Burlive oak	*Quercus macrocarpa x turbinella*	
Burr oak	*Quercus macrocarpa*	Mossy cup oak, blue oak
Butternut	*Juglans cinerea*	White walnut, oilnut
California black oak	*Quercus kelloggii*	Kellogg oak

Commonly known as	Latin name	Also known as
California live oak	*Quercus agrifolia*	Encina, coast live oak
California scrub oak	*Quercus dumosa*	Scrub oak
Californian wax myrtle	*Myrica californica*	Californian bayberry, Pacific wax myrtle
Canyon live oak	*Quercus chrysolepis*	Canyon oak, maul oak
Cherry plum	*Prunus cerasifera*	
Chestnut-leaved oak	*Quercus castaneifolia*	
Chestnut oak	*Quercus prinus*	Basket oak, rock oak
Chilgoza pine	*Pinus gerardiana*	Gerard's pine, Nepal nut pine
Chinese chestnut	*Castanea mollissima*	
Chinese cork oak	*Quercus variabilis*	Oriental oak
Chinese hazel	*Corylus chinensis*	
Chinese white pine	*Pinus armandii*	Armand's pine, David's pine
Chinkapin	*Castanea pumila*	Chinquapin
Chinkapin oak	*Quercus muehlenbergii*	Chestnut oak, yellow chestnut oak
Cinquefoils	*Potentilla* spp.	
Comfreys	*Symphytum* spp.	
Compton's oak	*Quercus lyrata x virginiana*	
Cork oak	*Quercus suber*	
Coulter's pine	*Pinus coulteri*	Big-cone pine
Crimson clover	*Trifolium incarnatum*	
Damson	*Prunus insititia*	
Digger pine	*Pinus sabiniana*	Ghost pine
Downy oak	*Quercus pubescens*	Pubescent oak
Dwarf chinkapin oak	*Quercus prinoides*	Chinkapin oak
Dwarf golden chinkapin	*Chrysolepis sempervirens*	Bush chinkapin, Sierra chinkapin
Eastern white pine	*Pinus strobus*	
Emory oak	*Quercus emoryi*	Western black oak
English live oak	*Quercus robur x turbinella*	
English oak	*Quercus robur*	Pedunculate oak
English white oak	*Quercus robur x alba*	
Filbert	*Corylus maxima*	
Gambel oak	*Quercus gambelii*	Shin oak
Ginkgo	*Ginkgo biloba*	Maidenhair tree
Ginseng	*Panax ginseng*	
Golden chinkapin	*Chrysolepis chrysophylla*	Golden chestnut, golden-leaved chestnut
Golden oak	*Quercus alnifolia*	
Goumi	*Elaeagnus multiflora*	
Green alder	*Alnus viridis*	
Ground-cover raspberries	*Rubus* spp.	
Hairy vetch	*Vicia villosa*	
Hawthorns	*Crataegus* spp.	

Commonly known as	Latin name	Also known as
Hazelnut	*Corylus avellana*	Hazel
Heartnut	*Juglans ailantifolia* var. *cordiformis*	
Hican	*Carya illinoinensis* x *Carya* spp.	
Himalayan sea buckthorn	*Hippophae salicifolia*	Willow-leaved sea buckthorn
Holly-leaved Grammont oak	*Quercus gramuntia*	
Holm oak	*Quercus ilex*	Holly oak
Huckleberry oak	*Quercus vacciniifolia*	
Hungarian oak	*Quercus frainetto*	Italian oak
Indian chestnut	*Castanopsis indica*	
Indian tree hazel	*Corylus jacquemontii*	
Interior live oak	*Quercus wislizeni*	Highland live oak
Israeli oak	*Quercus ithaburensis*	
Italian alder	*Alnus cordata*	
Japanese chestnut	*Castanea crenata*	
Japanese chinkapin	*Castanopsis cuspidata*	
Japanese emperor oak	*Quercus dentata*	Daimio oak
Japanese evergreen oak	*Quercus acuta*	
Japanese walnut	*Juglans ailantifolia*	
Jeffrey's pine	*Pinus jeffreyi*	
Johann's pinyon	*Pinus johannis*	
Kermes oak	*Quercus coccifera*	Grain oak
Konara oak	*Quercus glandulifera*	Gland-bearing oak
Korean barberry	*Berberis koreana*	
Korean pine	*Pinus korainensis*	Korean white pine, Korean nut pine, Chinese nut pine
Lady's mantle	*Alchemilla mollis*	
Lawson cypress	*Chamaecyparis lawsoniana*	
Lebanon oak	*Quercus libani*	
Libanerris oak	*Quercus* x *libanerris*	
Limber pine	*Pinus flexilis*	
Live oak	*Quercus virginiana*	Virginia live oak
Lucombe oak	*Quercus* x *hispanica*	Spanish oak
Lungwort	*Pulmonaria officinalis*	
Macedonian oak	*Quercus trojana*	
Maritime pine	*Pinus pinaster*	
Mexican blue oak	*Quercus oblongifolia*	Western live oak
Mexican pinyon	*Pinus cembroides*	Mexican stone pine, Mexican nut pine
Mexican white pine	*Pinus ayacahuite*	
Mockernut	*Carya tomentosa*	Big-bud hickory
Mongolian oak	*Quercus mongolica*	Japanese oak
Monkey puzzle	*Araucaria araucana*	Chile pine
Monterey cypress	*Cupressus macrocarpa*	

Commonly known as	Latin name	Also known as
Monterey pine	*Pinus radiata*	
Nelson's pinyon	*Pinus nelsonii*	
Northern California walnut	*Juglans hindsii*	
Northern pin oak	*Quercus ellipsoidalis*	Jack oak
Northern wax myrtle	*Myrica pennsylvanica*	Northern bayberry
Nutmeg hickory	*Carya myristiciformis*	
Nuttall's oak	*Quercus nuttallii*	
Oleaster	*Elaeagnus angustifolia*	
Ooti oak	*Quercus macrocarpa x muehlenbergii x robur*	
Oregon white oak	*Quercus garryana*	Garry oak, Oregon oak
Oriental white oak	*Quercus aliena*	
Overcup oak	*Quercus lyrata*	Swamp post oak
Paper-shell pinyon	*Pinus remota*	
Paraná pine	*Araucaria angustifolia*	
Parry pinyon	*Pinus cembroides quadrifolia (syns P. cembroides parrayana, P. parrayana, P. quadrifolia)*	Four-leaved nut pine
Pecan	*Carya illinoinensis*	
Periwinkles	*Vinca* spp.	
Pin oak	*Quercus palustris*	Spanish oak, swamp oak, swamp Spanish oak
Pince's pinyon	*Pinus pinceana*	
Pinyon pine	*Pinus cembroides edulis* (syn. *P. edulis*)	Piñon pine, Colorado piñon, nut pine, two-leaved nut pine, two-needle pinyon, silver pine, Rocky Mountain nut pine
Portuguese oak	*Quercus faginea*	
Post oak	*Quercus stellata*	Iron oak
Pyrenean oak	*Quercus pyrenaica*	Spanish oak
Red alder	*Alnus rubra*	
Red oak	*Quercus rubra*	Northern red oak, American red oak
Red pine	*Pinus resinosa*	Norway pine
Robata oak	*Quercus robur x lobata*	
Sandpaper oak	*Quercus pungens*	
Sawtooth oak	*Quercus acutissima*	Korean oak
Scarlet oak	*Quercus coccinea*	Spanish oak
Schuette's oak	*Quercus x schuettes*	
Schumard oak	*Quercus shumardii*	Schneck oak
Sea buckthorn	*Hippophae rhamnoides*	
Sessile oak	*Quercus petraea*	Durmast oak
Shagbark hickory	*Carya ovata*	
Shellbark hickory	*Carya laciniosa*	
Shingle oak	*Quercus imbricaria*	Laurel oak

Commonly known as	Latin name	Also known as
Siberian nut pine	*Pinus sibirica* (syn. *P. cembra sibirica*)	Siberian pine, Siberian cedar, cedar pine, Russian cedar
Siberian pea tree	*Caragana arborescens*	
Siberian purslane	*Claytonia sibirica*	
Silverberry	*Elaeagnus commutata*	
Singleleaf pinyon	*Pinus cembroides monophylla* (syn. *P. monophylla*)	
Sitka alder	*Alnus sinuata*	
Southern red oak	*Quercus falcata*	Swamp red oak
Stone pine	*Pinus pinea*	Umbrella pine, Italian stone pine
Sugar pine	*Pinus lambertiana*	Lambert pine
Swamp chestnut oak	*Quercus michauxii*	Cow oak, basket oak
Swamp white oak	*Quercus bicolor*	White oak
Sweet chestnut	*Castanea sativa*	European sweet chestnut
Sycamore	*Acer pseudoplatanus*	
Tibetan chinkapin	*Castanopsis tibetana*	
Torrey pine	*Pinus torreyana*	Soledad pine
Trazels	*Corylus avellana* x *Corylus* spp.	
Turkey oak	*Quercus cerris*	
Turkish hazel	*Corylus colurna*	
Ubame oak	*Quercus phillyreoides*	Peach oak, pin oak
Valley oak	*Quercus lobata*	California white oak
Vallonea oak	*Quercus ithaburensis* subsp. *macrolepis*	Camata, Valonea oak
Walnut	*Juglans regia*	English walnut, Persian walnut
Water hickory	Carya aquatic	
Water oak	*Quercus nigra*	Possum oak
Wavyleaf oak	*Quercus undulata*	
Wax myrtle	*Myrica cerifera*	Bayberry
White oak	*Quercus alba*	Stave oak, Quebec oak, American white oak
Whitebark pine	*Pinus albicaulis*	
Wild garlic	*Allium ursinum*	
Willow oak	*Quercus phellos*	
Willows	*Salix* spp.	
Yellowhorn	*Xanthoceras sorbifolium*	Shiny-leaf yellowhorn, northern macademia

Resources

The following resources are divided by region: Europe (pages 304-306), North America (pages 306-308) and Australasia (pages 308-309). The bibliography is separate on page 309.

A note about drying equipment: for home-scale drying, small dehydrators can be sufficient. There are many types readily available, but among the best are 'Excalibur' dehydrators. Larger commercial dehydrators are made in many countries; alternatively, it is quite easy to make your own nut dryer (see Chapter 3, page 72).

Europe

Nut tree nurseries

Agroforestry Research Trust (UK)
46 Hunters Moon, Dartington, Totnes, Devon, TQ9 6JT
Tel: + 44 (0)1803 840776
www.agroforestry.co.uk
mail@agroforestry.co.uk

Ahornblatt GmbH (Germany)
Untere Zahlbacher Straße 1a, 55131 Mainz-Zahlbach, Germany
Tel: + 49 (0)6131 72354
http://ahornblatt-garten.de
Nachricht@Ahornblatt-Garten.de

Baumschule Anton Schott (Germany)
Steuernbergstraße 2, 79361 Leiselheim, Germany
Tel: +49 (0)7642 5859
www.nussspezialist.de/
info@nussspezialist.de

Baumschule H Brillinger (Austria)
Gartenstraße 1, 4616 Weißkirchen bei Wels, Oberösterreich, Austria
Tel: +43 (0)7243 56120
www.baumschule-brillinger.at
office@brillinger.at

Baumschule Horstmann (Germany)
Nursery Horstmann GmbH & Co. KG, Bergstraße 5, 25582 Hohenaspe, Germany
Tel: +49 (0)4892 8993400
www.baumschule-horstmann.de
info@baumschule-horstmann.de

Baumschule Mayer GbR (Germany)
Mühläcker 12, 97990 Weikersheim-Elpersheim, Germany
Tel: +49 (0)7934 991100
www.baumschule-mayer.de
info@baumschule-mayer.de

Baumschule New Garden (Germany)
Prozessionsweg 62, 46325 BorkenWeseke, Germany
Tel: +49 (0)2862 700207
www.baumschule-newgarden.de
info@new-garden.de

Baumschule Pernerstorfer (Austria)
Waldviertler Baumschule
Hans Pernerstorfer, Kremser Straße 11, A-3542 Gföhl, Austria
Tel: +43 (0)2716 6456
www.pernerstorfer.at
baumschule.pernerstorfer@aon.at

Blackmoor Nurseries (UK)
Blackmoor, Nr Liss, Hampshire, GU33 6BS
Tel: +44 (0)1420 477978
www.blackmoor.co.uk
sales@blackmoor.co.uk

Boomkwekerij De Acht Plagen (Netherlands)
Middelweg west 203, 9079 MK Sint Jacobiparochie, Netherlands
Tel: +31 (0)629 174239
http://achtplagennuts.nl
achtplagennuts@12move.nl

Cool Temperate (UK)
Newtons Lane, Cossall, Notts, NG16 2YH
Tel: + 44 (0)115 916 2673
www.cooltemperate.co.uk

De Smallekamp (Netherlands)
Hardenbrinkweg 24, 8071 SM Nunspeet, Netherlands
Tel: +31 (0)653 817890
www.desmallekamp.nl
info@desmallekamp.nl

Deacon's Nursery (UK)
Moor View, Godshill, Isle of Wight, PO38 3HW
Tel: +44 (0)1983 840750 or (0)1983 522243
www.deaconsnurseryfruits.co.uk
info@deaconsnurseryfruits.co.uk

Der Sängerhof (Germany)
Germany
Tel: +49 (0)2225 993170
www.gartenwebshop.eu

Eggert Baumschulen (Germany)
Baumschulenweg 2, 25594 Vaale, Germany
Tel: +49 (0)4827 932627
www.eggert-baumschulen.de
kontakt@eggert-baumschulen.de

Frank P Matthews Ltd (UK)
Berrington Ct, Tenbury Wells, Worcs, WR15 8TH
Tel: + 44 (0)1584 810214
www.frankpmatthews.com
enquiries@fpmatthews.co.uk

Frédéric Cochet (France)
07200 Aubenas, 48 chemin de St Pierre, France
Tel: +33 (0)4 75 35 91 90
www.cochetfrederic.com
contact@cochetfrederic.com

Fruit and Nut (Ireland)
The Sustainability Institute, Cooloughra, Ballinrobe Rd,
Westport, Co. Mayo, Ireland
Tel: +353 (0)87 6714075
www.fruitandnut.ie
office@fruitandnut.ie

Gruener Garden Shop (Germany)
Bielefelder Str. 202, 32758 Detmold, Germany
Tel: +49 (0)5231 3077334
www.gruener-garten-shop.de
info@gruener-garten-shop.de

Haas & Haas (Austria)
A-2063 Zwingendorf 39, Austria
Tel: +43 (0)2527 324
http://walnussbaum.at
office@garten-haas.at

Keepers Nursery (UK)
Gallants Ct, East Farleigh, Maidstone, Kent, ME15 0LE
Tel: +44 (0)1622 326465
www.keepers-nursery.co.uk
helpdesk@keepers-nursery.co.uk

Kwekerij Westhof (Netherlands)
Westhofsezandweg 3, 4444 SM 's-Heer Abtskerke (Zld.),
Netherlands

Tel: +31 (0)113 561219
http://walnoten.nl
westhof@walnoten.nl

La Pépinière du Bosc (France)
Route de Lodève, 34700 St Privat, France
Tel: +33 (0)6 61 65 34 20
www.pepinieredubosc.fr
contact@pepinieredubosc.fr

Otter Farm (UK)
http://shop.otterfarm.co.uk
info@otterfarm.co.uk

Pépinière Noyers Lalanne (France)
Benoit Geneau de Lamarliere, Navail 47180 Ste Bazeille /
Marmande, France
Tel: +33 (0)6 85 10 04 82
www.pepinieres-noyers-lalanne.com
contact@pepinieres-noyers-lalanne.com

Pépinières Coulié (France)
Le Sorpt, 19600 Chasteaux, France
Tel: +33 (0)5 55 85 34 21
www.coulie.com

Pépinières du Pondaillan (Linard) (France)
Rue du Pondaillan, 46200 Souillac, France
Tel: +33 (0)5 65 37 83 17
www.l-q-p.com

Pépinières Lafitte (France)
Quartier Greciette, 64240 Mendionde, France
Tel: +33 (0)5 59 29 62 54
www.lafitte.net/fr/pepinieres-lafitte
info@lafitte.net

Pépinières Louis Gauthier (France)
187 chemin des Paluds, 13670 St Andiol, France
Tel: +33 (0)4 90 95 14 33
www.pepinieres-gauthier.fr
louis.gauthier@9online.fr

Pépinières Mouraud (France)
46200 Pinsac, France
Tel: +33 (0)6 08 60 00 47
www.mouraud.fr
contact@mouraud.fr

Pépinières Payre (France)
205 rue des Terreaux, 38470 L'Albenc, France
Tel: +33 (0)4 76 64 76 60
www.pepiniere-payre-lalbenc.fr
pepiniere.payre@wanadoo.fr

Pépinières Pepinoix (France)
La Thivolière, 38210 Polienas, France
Tel: +33 (0)4 76 07 22 60

www.pepinoix.com
sebastien.desbrus@pepinoix.com

Reads Nursery (UK)
Douglas Farm, Falcon Lane, Ditchingham, Bungay, Suffolk, NR35 2JG
Tel: +44 (0)1986 895555
www.readsnursery.co.uk

SCEA les Coteaux de Boutau (France)
Labarthe, 33190 Camiran, France
Tel: +33 (0)5 56 61 58 20
sceaboutau@wanadoo.fr

Small-scale equipment

Feucht-Obsttechnik (Germany)
Europastr. 16, D-71576 Erbstetten, Germany
Tel: +49 (0)7191 64195
www.feucht-obsttechnik.de
juergen.feucht@feucht-obsttechnik.de
Supplies: Nut-cracking machines.

Mladen Gudan
(Republic of Croatia)
Mehanovizija doo, Franje Jurinca 105, 10310 Ivanić Grad, Republic of Croatia
mehanovizija@net.hr
Supplies: Cronut nut-cracking machine and Nut Wizards.

Piteba (Netherlands)
www.piteba.com
Supplies: Piteba oil press.

Turco Bazaar (Turkey)
Söğütlüçeşme Cad. Abdullah Uzlar İş Merkezi, No: 92-94/49 Kadıköy/İstanbul, Turkey
Tel: +90 541 825 8498
www.turcobazaar.com
info@turcobazaar.com
Supplies: Nutcrackers and separators.

Walnut Wizard (UK)
www.walnutwizard.com
Walnut Nut Wizard manufacturer with links to European distributors.

Larger-scale equipment

AMB Rousset (France)
281 Impasse du Tilleul, 38470 Beaulieu, France
Tel: +33 (0)4 76 36 73 73
www.amb-rousset.com
Supplies: Harvesting and processing equipment.

Arnaud et Blanc (France)
3 avenue de Romans, 38160 St Marcellin, France
Tel: +33 (0)8 90 39 11 76
www.societe.com/societe/arnaud-et-blanc-447671645.html
Supplies: Harvesting and processing equipment.

Aubert et Cie (France)
9 avenue de l'Industrie, 24660 Coulounieix Chamiers, France
Tel: +33 (0)8 90 39 11 32
www.societe.com/societe/etablissements-aubert-et-compagnie-581980158.html
Supplies: Processing equipment.

Bobard Jeune (France)
17 rue de Reon, 21200 Beaune, France
Tel: +33 (0)8 90 39 09 88
www.societe.com/societe/etablissements-bobard-jeune-317356632.html
Supplies: Mechanical harvesting equipment.

Constructions Mécaniques du Villeneuvois (CMV) (France)
Z.I. La Barbière, rue Paul Langevin, 47300 Villeneuve/Lot, France
Tel: +33 (0)5 53 70 52 89
www.cmv47.com
Supplies: Hand-operated nut harvesters.

Feucht-Obsttechnik (Germany)
Europastr. 16, D-71576 Erbstetten, Germany
Tel: +49 (0)7191 64195
www.feucht-obsttechnik.de
juergen.feucht@feucht-obsttechnik.de
Supplies: Harvesting and processing equipment.

Giampi (Italy)
Loc. Capannelle, Z.Art.le, 01030 Carbognano (VT), Italy
Tel: +39 (0)761 572384
www.giampimacchineagricole.com
giampimacchineagricole@gmail.com
Supplies: Harvesting equipment.

Monchiero (Italy)
40 Strada Crociera Burdina, 12042 Pollenzo di Bra (CN), Italy
Tel: +39 (0)172 458126
www.monchiero.com
Supplies: Harvesting equipment.

Terreco (France)
Terreco SARL Euroleix, Le Champ des Vergnes, 19240 Allassac, France

Tel: +33 (0)5 55 85 05 94
www.arboriculture-viticulture.pro
contact@terreco.pro
Supplies: Harvesting and processing equipment.

Organizations

CTIFL (France)
22 rue Bergère, 75009 Paris, France
Tel: +33 (0)1 47 70 16 93
www.ctifl.fr
info@ctifl.fr
A non-profit technical centre aimed at improving the level of expertise in the fruit and vegetable industry. Publishes excellent books (in French) on individual nut crops.

IAMZ-CIHEAM (Spain)
Mediterranean Agronomic Institute of Zaragoza, Avenida de Montanana 1005 50059, Zaragoza, Spain
Tel: +34 976 716000
www.iamz.ciheam.org
iamz@iamz.ciheam.org
The CIHEAM (International Centre for Advanced Mediterranean Agronomic Studies) hosts a Mediterranean-based nut research network and publishes its findings in the Nucis newsletter.

Kentish Cobnuts Association (UK)
Alexander Hunt, Apple Trees, Comp Lane, St Mary's Platt, Nr Sevenoaks, Kent, TN15 8NR
Tel: +44 (0)1732 882734
http://kentishcobnutsassociation.org.uk
info@kentishcobnutsassociation.org.uk
An association of hazel growers that runs courses on pruning and plant management and publishes a newsletter.

North America

Nut tree nurseries

Aaron's Farm
PO Box 800, Sumner, GA 31789, USA
Tel: +1 888 730 4032
www.aaronsfarm.com
sales@aaronsfarm.com

Arbornaut Nursery
Denman Island, BC, Canada
www.arbornautnursery.com
arbornautnursery@gmail.com

Bay Laurel Nursery
2500 El Camino Real, Atascadero, CA 93422, USA
Tel: +1 805 466 3449
https://baylaurelnursery.com

Burnt Ridge Nursery
432 Burnt Ridge Rd, Onalaska, WA 98570, USA
Tel: +1 360 985 287
www.burntridgenursery.com
mail@burntridgenursery.com

Chestnut Hill Outdoors
15105 NW 94 Ave Alachua, FL 32615, USA
Tel: +1 855 386 7826
www.chestnuthilloutdoors.com
chestnuthilloutdoors@gmail.com

Edible Landscaping
361 Spirit Ridge Lane, Afton, VA 22920, USA
Tel: +1 434 361 9134
http://ediblelandscaping.com
info@ediblelandscaping.com

Empire Chestnut Company
3276 Empire Rd SW, Carrollton, OH 44615-9515, USA
http://empirechestnut.com
empirechestnut@gmail.com

England's Orchard & Nursery
2338 Highway 2004, Mckee, KY 40447-8342, USA
Tel: +1 606 965 2228
http://nuttrees.net
nuttrees@prtcnet.org

Grimo Nut Nursery
979 Lakeshore Rd, R.R.3, Niagara-on-the-Lake, Ontario, L0S 1J0, Canada
Tel: +1 905 934 6887
www.grimonut.com

Hardy Fruit Tree Nursery
Box 5754, Sainte-Julienne, Quebec, J0K 2T0, Canada
Tel: +1 514 418 4109
http://hardyfruittrees.ca

Harvest Nursery
10470 NE 6th Drive, Portland, OR 97211, USA
Tel: +1 503 332 0913
http://harvestnursery.com

Nolin River Nut Tree Nursery
797 Port Wooden Rd, Upton, Kentucky 42784, USA
Tel: +1 270 369 8551
http://nolinnursery.com
john.brittain@windstream.net

Nutcracker Nursery
320, rang Rivière Sud-Ouest, Maskinongé (Québec), J0K 1N0, Canada
Tel: +1 819 386 4834
www.nutcrackernursery.com
info@cassenoisettepepiniere.com

Oikos Tree Crops
PO Box 19425, Kalamazoo, MI 49019-0425, USA
Tel: +1 269 624 6233
http://oikostreecrops.com

One Green World
6469 SE 134th Ave, Portland, OR 97236, USA
Tel: +1 877 353 4028
https://onegreenworld.com

Pépinière Casse-Noisette
www.cassenoisettepepiniere.com
Contact details as for **Nutcracker Nursery**

Raintree Nursery
391 Butts Rd, Morton, WA 98356, USA
Tel: +1 800 391 8892
www.raintreenursery.com

Rhora's Nut Farm and Nursery
Charles A. Rhora, R.R. #1, 33083 Wills Rd, Wainfleet,
Ontario, L0S 1V0, Canada
Tel: +1 905 899 3508
www.nuttrees.com
rhoras@nuttrees.com

Rolling River Nursery
PO Box 332, Orleans, CA 95556, USA
Tel: +1 530 627 3120
www.rollingrivernursery.com
info@rollingrivernursery.com

St. Lawrence Nurseries
PO Box 957, Potsdam, NY 13676, USA
Tel: +1 315 261 1925
www.sln.potsdam.ny.us
connor@sln.potsdam.ny.us

Stark Bro's
PO Box 1800, Louisiana, MO 63353, USA
Tel: +1 800 325 4180
www.starkbros.com
info@starkbros.com

Ty Ty Nursery
PO Box 130, Ty Ty, GA 31795-0130, USA
Tel: +1 888 758 2252
www.tytyga.com
sales@tytyga.com

Willis Orchard Co.
200 McCormick Rd, Cartersville, GA 30120, USA
Tel: +1 866 586 6283
www.willisorchards.com
orders@willisorchards.com

Small-scale equipment

Badgersett
Badgersett Research Corporation, 18606 Deer Rd,
Canton, MN 55922, USA
www.badgersett.com
info@badgersett.com
Supplies: Chestnut-peeling pliers.

Bag-a-Nut
10601 Theresa Drive, Jacksonville, FL 32246, USA
Tel: +1 800 940 2688
http://baganut.com
Supplies: Nut harvesters.

DAVEBILT co.
410 Soda Bay Rd, Lakeport, CA 95453, USA
Tel: +1 707 263 5270
www.davebilt.com
contact@davebilt.com
Supplies: Nutcracker and Little Davey nut picker.

Grimo Nut Nursery
979 Lakeshore Rd, R.R.3, Niagara-on-the-Lake, Ontario,
L0S 1JO, Canada
Tel: +1 905 934 6887
www.grimonut.com
Supplies: Nut Wizards and Little Davey nut picker.

Kenkel, Inc.
1233 Champlaine Ct., Schaumburg, IL 60193, USA
Tel: +1 847 923 1987
www.kenkelnutcracker.com
mail@kenkelnutcracker.com
Supplies: Heavy-duty nutcracker.

Lawn Gardening Tools
2678 Reeves St, Dothan, Alabama 36303, USA
Tel: +1 334 246 5798
http://lawn-gardening-tools.com
Supplies: Nut harvesters and nutcrackers.

Nut Wizard
Seeds and Such, Inc., 1105 R St, PO Box 81, Bedford, IN
47421, USA
Tel: +1 888 321 9445
http://nutwizard.com
nutwizard@hpcisp.com
Supplies: Nut Wizards.

Universal Nutcracker
1694 W. 6th Ave, Junction City, OR 97448, USA
Tel: +1 541 954 8588
http://universalnutcracker.com
info@universalnutcracker.com
Supplies: Heavy-duty, high-volume nutcrackers.

The World's Best Nutcracker
Reflections, PO Box 1163, Rough & Ready, CA 95975, USA
Tel: +1 530 273 9378
www.theworldsbestnutcracker.com
nutcracker@photomagnets.com
Supplies: Nutcrackers, push harvesters, oil presses.

Larger-scale equipment

Bag-a-Nut
10601 Theresa Drive, Jacksonville, FL 32246, USA
Tel: +1 800 940 2688
http://baganut.com
Supplies: Nut harvesters.

Flory
Flory Industries, 4737 Toomes Road, PO Box 908
Salida, CA 95368, USA
Tel: +1 800 662 6677
http://goflory.com/
sales@goflory.com
Supplies: Harvesting equipment.

Lakewood Processing Machinery
875 Brooks Ave, Holland, MI 49423, USA
Tel: +1 800 366 6705
http://lakewoodpm.com/
info@lakewoodpm.com
Supplies: Nut-sizing machines

Orchard Machinery Corporation (OMC)
2700 Colusa Highway, Yuba City, CA 95993, USA
Tel: +1 530 673 2822
www.shakermaker.com
Supplies: Harvesting equipment.

Orchard-Rite
1615 W Ahtanum Rd, Yakima, WA 98903, USA
Tel: +1 509 834 2029 (ext 620)
www.orchard-rite.com/
sales@orchard-rite.com
Supplies: Harvesting equipment.

Savage
1020 North Industrial Road, Madill, Oklahoma 73446,
USA
Tel: +1 866 572 8243
www.savageequipment.com/
info@savageequipment.com
Supplies: Harvesting and processing equipment.

Universal Nut Cracker
1694 W. 6th Ave., Junction City, OR 97448, USA
Tel: +1 541 954 8588
http://universalnutcracker.com

info@universalnutcracker.com
Supplies: Nut-cracking and separating equipment.

Weiss McNair
100 Loren Avenue, Chico, CA 95928, USA
Tel: +1 530 891 6214
www.weissmcnair.com/
Supplies: Harvesting equipment.

Organisations

Badgersett
Badgersett Research Corporation, 18606 Deer Road,
Canton, MN 55922, USA
www.badgersett.com
info@badgersett.com
Private breeding organisation developing hardy hybrid
populations of hazel.

California Oaks
428 13th Street, Suite 10A, Oakland, CA 94612, USA
Tel: +1 510 763 0282
www.californiaoaks.org/ExtAssets/acorns_and_eatem.
pdf
Dedicated to California oak and wildlife habitat sustaina-
bilty. The website includes a number of oak-related books
available as free downloads, including *Acorns and Eat 'em*
by Suellen Ocean, an excellent book on eating acorns.

Northern Nut Growers Association (NNGA)
www.nutgrowing.org
Nut-growing queries: email Tucker Hill at tuckerh@epix.
net
Membership queries: email Grant Glatt at grant_40@
juno.com
Non-profit organisation, with members throughout the
USA and Europe, founded in 1910 to share information on
nut tree growing. The website contains plentiful useful
resources, recipes and articles on nut nutrition.

Australia and New Zealand

Nut tree nurseries

Australian Gourmet Hazelnuts
Tel: +61 (0)2 6372 3224
www.gourmethazelnuts.com.au
ausnuts@gourmethazelnuts.com.au

Daleys
PO Box 154, Kyogle NSW 2474, Australia
www.daleysfruit.com.au

Edible Garden
889 Ashhurst-Bunnythorpe Road, RD10 Palmerston

North, 4470, New Zealand
Tel +64 (0)6 326 7313
http://ediblegarden.nz

Fleming's Nurseries
PO Box 1, Monbulk Victoria 3793,
Australia
Tel: +61 (0)3 9756 6105
www.flemings.com.au
mail@flemings.com.au

Hazelnut Nursery Propagators
PO Box 364, Gembrook, Victoria
3783, Australia
Tel: +61 (0)3 5968 1092
www.hazelnuts.com.au
info@hazelnuts.com.au

The Merry Nutfarm
Justine and Anthony Merry, 9 Priors
Rd, The Patch, Victoria 3792, Australia
Tel: +61 (0)3 9756 7998
www.themerrynutfarm.com.au
merrynutfarm@bigpond.com

Orchard Fruits Nursery
PO Box 491, Loxton, SA 5333, Australia
Tel: +61 (0)8 8584 5544
http://pippos.com
nursery@pippos.com

Perry's Fruit & Nut Nursery
Kangarilla Rd, Mclaren Flat, SA 5171,
Australia
Tel: +61 (0)8 8383 0268
www.perrysfruitnursery.com.au

Rossmount Nursery
2 Burns Rd, Goomboorian Via
Gympie, Queensland 4570, Australia
Tel: +61 (0)7 5483 3734
www.rossmount.com.au
admin@rossmount.com.au

Southern Woods Plant Nursery
1133 Main South Rd, Templeton,
Christchurch 8042, New Zealand
Tel: +64 (0)800 800 352
www.southernwoods.co.nz
info@southernwoods.co.nz

Waimea Nurseries
www.waimeanurseries.co.nz
Distribute through a network of
garden centres.

Wairata Hazels
Tel: +64 (0)7 315 7763
www.wairatahazels.co.nz
murray@wairatahazels.co.nz

Small-scale equipment

Australian Nut Harvesters
EMARA, 437 Ironcliffe Rd, Penguin, Tasmania 7316, Australia
Tel: +61 (0)407 847 170
www.nutharvester.com.au
john.pethybridge@gmail.com
Supply: Nut Wizards and hand-cranked nutcracker.

Larger-scale equipment

Flory
See North America, page 308.

Monchiero Australia
Tel: +61 (0)439 4049 78
http://www.nutharvesters.com/
aljomorgan@hotmail.com
Supply: Italian-made Monchiero harvesting equipment.

Orchard Machinery Corporation (OMC)
See North America, page 308.

Orchard-Rite Australia
695 Elizabeth Ave, Wagga Wagga, NSW 2650, Australia
Tel: +61 (0)2 6922 8689
www.orchard-rite-australia.com.au/
enquiries@orchard-rite-australia.com.au
Supply: Harvesting equipment.

Savage
See North America, page 308.

Organizations

Australian Nut Industry Council
Chasely Ross, Executive Officer, 42 Simpsons Rd,
Currumbin Waters, Queensland 4223, Australia
Tel: +61 (0)409 707 806
www.nutindustry.org.au
exec@nutindustry.org.au
Publishes the *Australian Nutgrower* journal.

New Zealand Chestnut Council
David Klinac, 10 June Place, Hamilton 3216, New Zealand
Tel: +64 (0)7 856 9321
www.nzcc.org.nz
dklinac@xtra.co.nz
Has developed chestnut-peeling machines.

Bibliography

Avanzato, D. (ed) (2014). *Following Walnut Footprints* (Juglans regia L.): *Cultivation and culture, folklore and history, traditions and uses. Scripta Horticulturae* 17. International Society for Horticultural Science (ISHS): Leueven, Belgium.

Breisch, H. (1995). *Châtaignes et marrons*. Le Centre technique interprofessionnel des fruits et légumes (CTIFL): Paris, France.

Chenoweth, B. (1995). *Black Walnut: The history, use, & unrealized potential of a unique American*. Sagamore Publishing: Illinois, USA.

Crawford, M. (1995). *Chestnuts, Production and Culture*. Agroforestry Research Trust: Totnes, Devon, UK.

Crawford, M. (1995). *Hazelnuts: Production and culture*. Agroforestry Research Trust: Totnes, Devon, UK.

Crawford, M. (1996). *Walnuts: Production and culture*. Agroforestry Research Trust: Totnes, Devon, UK.

Crawford, M. (2010). *Creating a Forest Garden: Working with nature to grow edible crops*. Green Books: Dartington, Devon, UK.

Crawford, M. (2015). *Trees for Gardens, Orchards and Permaculture*. Permanent Publications: East Meon, Hampshire, UK.

Douglas, S. and De J. Hart, R. (1985). *Forest Farming: Towards a solution to problems of world hunger and conservation*. 3rd edn. ITDG Publishing: Bradford, West Yorkshire, UK.

Duke, J. A. (1989). *CRC Handbook of Nuts*. CRC Press: Florida, USA.

Farris, C. W. (2000). *The Hazel Tree*. Northern Nut Growers Association (NNGA): Connecticut, USA.

Fulbright, D. W. (ed) (2003). *A Guide to Nut Tree Culture in North America Vol 1*. Northern Nut Growers Association (NNGA): Connecticut, USA.

Garner, R. J. (1993). *The Grafter's Handbook*. Cassell: London, UK.

Germain, E. et al. (1999). *Le Noyer*. CTIFL: Paris, France.

Germain, E. et al. (2004). *Le Noisetier*. CTIFL: Paris, France.

Grasselly, C. et al. (1997). *L'Amandier*. Le Centre technique interprofessionnel des fruits et légumes (CTIFL): Paris, France.

Grimo, E. (2011). *Nut Tree Ontario: A practical guide*. Society of Ontario Nut Growers (SONG): Ontario, USA.

Howes, F. N. (1953). *Nuts: Their production and everyday uses*. 2nd edn. Faber and Faber: London, UK.

Jaynes, R. A. (ed) (1979). *Nut Tree Culture in North America*. Northern Nut Growers Association (NNGA): Connecticut, USA.

Micke, W. (ed) (1996). *Almond Production Manual*. Publication 3364. University of California, Division of Agriculture and Natural Resources (DANR): California, USA.

Mitra, S. K. et al. (eds) (1991). *Temperate Fruits*. Horticulture and Allied Publishers: Calcutta, India.

Moore, J. N. and Ballington Jr., J. R. (eds) (1990). *Genetic Resources of Temperate Fruit and Nut Crops, Vol 2. Acta Horticulturae* 290. International Society for Horticultural Science (ISHS): Leueven, Belgium.

Ocean, S. (1995). *Acorns and Eat 'em: A how-to vegetarian cookbook*. Ocean-Hose: California, USA.

Paterson, R. (1993). *Use of Trees by Livestock: Quercus*. Natural Resources Institute (NRI): Chatham Maritime, Kent, UK.

Ramos, D. E. (ed) (1998). *Walnut Production Manual*. University of California, Division of Agriculture and Natural Resources (DANR): California, USA.

Reed, C. A. and Davidson, J. (1954). *The Improved Nut Trees of North America and How to Grow Them*. Devin-Adair: New York, USA.

Sauvezon, R. et al. (2000). *Châtaignes et châtaigniers*. Édisud: St-Rémy-de-Provence, France.

Woodroof, J. G. (1979). *Tree Nuts: Production, processing, products*. 2nd edn. AVI Publishing: Connecticut, USA.

Online publications

Oregon State University (OSU) has many useful publications on growing hazelnuts: http://extension.oregonstate.edu/yamhill/hazelnuts-filberts

Photo credits

1-2, 5 (bottom right, bottom middle), 6, 15, 16-17, 18, 41, 44-45, 64-65, 83, 235 and 276 © **Joanna Brown**.

5 (top), 7, 8, 9 , 22, 28, 30, 47, 53, 62, 96, 148 and 158 via **Shutterstock**.

262: **United States Department of Agricuture** (USDA)

Via Wikipedia Commons:

48: Kapková závlaha, napojení potrubí na rozvod vody s kohoutem.JPG, **Juandev**

54: Eastern Grey Squirrel in St James's Park, London - Nov 2006 edit.jpg, **Diliff**

57: Corvus corone mit einer Walnuss 02.JPG, **4028mdk09**

89: Flor Almendro.JPG, **Mario modesto**

90: Prunus dulcis.jpg, **Mike Stephenson**, Image: mickstephenson.photoshelter.com

92: Anarsia lineatella (9380798139).jpg, **Donald Hobern**

93: Taphrina deformans 2.jpg, **Giancarlo Dessi**

116: Staphylea trifolia SCA-3465.jpg, **R. A. Nonenmacher**

117 (top): Staphylea trifolia SCA-3489.jpg, **R. A. Nonenmacher**

126 (right): Juglans cinerea female.jpg, Steven Katovich, USDA Forest Service

126 (left): Juglans inerea male.jpg, **Steven Katovich**, USDA Forest Service

136: Ginkgo biloba - male flower.JPG, **Marcin Kolasiński**

137: Ginkgo biloba - female flower.JPG, **Marcin Kolasiński**

138 (top left): Ginkgo biloba 007.jpg, **H. Zell**

140: Villa oliva, cascatelle, ginkgo biloba 01.JPG , **sailko**

141: Ginkgo biloba 004.JPG, **H. Zell**

143: Chrysolepis chrysophylla Huckleberry BRP 2.jpg, **Mike Linksvayer**

144 (right): Chrysolepis sempervirens.jpg, **jkirkhart35**

144 (left): Castanopsis sieboldii (Makino) Hatus. — スダジ イ by urasimaru.jpg, **urasimaru**

145: Castanopsis sieboldii nuts02.jpg, **Hamachidori**

150 (left): Corylus avellana 0007.JPG, **H. Zell**

153 (left): 3494 – Cydia latiferreana – Filbertworm Moth (21466795848).jpg, **Andy Reago** / **Chrissy McClarren**

153 (right): Anisogramma anomala.jpg, **Joseph O'Brien**, USDA Forest Service

172: Kornik Arboretum orzesznik siedmiolistkowy.jpg, **Andrzej Otrębski**

174: Shagbark Hickory trunks Basking Ridge NJ.jpg, **John B**

176: Hickory nuts 6060.JPG, **Abrahami**

194 (top left): Quercus cerris acorn BG 2.jpg, **Dimìtar Nàydenov**

194 (top right): QUERCUS COCCIFERA - AGUDA - IB-318 (Garric).JPG, **Isidre Blanc**

194 (centre left): Quercus ilex, Livorno.JPG , **Lucarelli**

194 (centre right): Quercus macrocarpa (5107489209).jpg, **Matt Lavin**

194 (bottom left): Acorns - Quercus petraea.jpg, **T137**

194 (bottom right): Quercus robur 002.JPG, **Llez**

202: Pecan tree canopy zilker.jpg, **Larry D. Moore** / US Department of Agriculture

203: Carya illinoinensis foliagenuts.jpg, **Brad Haire**, University of Georgia, USA

205: Carya illinoinensis catkins.jpg, **Clemson University** – USDA Cooperative Extension Slide Series, USA.

207 (top): Pecanscab.jpg, **Clemson University** – USDA Cooperative Extension Slide Series, Bugwood.org

207 (bottom): 20111209-NRCS-LSC-0029 - Flickr - USDA-gov.jpg, **Lance Cheung**

212: Pinus pinea Doñana 1.jpg, **Nacho Pintos**

217: Cronartium_ribicola_bialowieza_4_beentree.jpg, **Beentree**

218: PINUS PINEA - AGUDA - IB-109 (Pi_pinyer).JPG, **Isidre Blanc**

233: Cydia splendana01.jpg, **Kurt Kulac** kulac@gmx.at

234: 20150511Castanea sativa7.jpg, **AnRo0002**

261: 2015-05 ND-41 Baumhasel E-Leithe.jpg, **Yvonne Bentele**

265: Noyers-ferme_15.JPG, © **Traumrune** / Wikimedia Commons / CC-BY-3.0

272 (left): Cydia pomonella, Lodz(Poland)01(js).jpg, **Jerzy Strzelecki**

292: Xanthoceras_sorbifolium1.jpg, **Athenchen**.

Index

Page numbers in **bold** indicate the main entry for a nut; page numbers in *italic* refer to illustrations.